Lecture Notes in Physics

The Lecture Notes in Physics

The series Lecture Notes in Physics (LNP), founded in 1969, reports new developments in physics research and teaching – quickly and informally, but with a high quality and the explicit aim to summarize and communicate current knowledge in an accessible way. Books published in this series are conceived as bridging material between advanced graduate textbooks and the forefront of research and to serve three purposes:

- to be a compact and modern up-to-date source of reference on a well-defined topic

- to serve as an accessible introduction to the field to postgraduate students and nonspecialist researchers from related areas

- to be a source of advanced teaching material for specialized seminars, courses and schools

Both monographs and multi-author volumes will be considered for publication. Edited volumes should, however, consist of a very limited number of contributions only. Proceedings will not be considered for LNP.

Volumes published in LNP are disseminated both in print and in electronic formats, the electronic archive being available at springerlink.com. The series content is indexed, abstracted and referenced by many abstracting and information services, bibliographic networks, subscription agencies, library networks, and consortia.

Proposals should be sent to a member of the Editorial Board, or directly to the managing editor at Springer:

Christian Caron
Springer Heidelberg
Physics Editorial Department I
Tiergartenstrasse 17
69121 Heidelberg / Germany
christian.caron@springer.com

John B. Parkinson · Damian J.J. Farnell

An Introduction to Quantum Spin Systems

 Springer

John B. Parkinson
University of Manchester
School of Mathematics
Oxford Road
M13 9PL Manchester
United Kingdom
J.B.Parkinson@manchester.ac.uk

Damian J.J. Farnell
University of Manchester
Jean McFarlane Building
School of Community-Based
Medicine
Health Methodology Research
Group
University Place
M13 9PL Manchester
United Kingdom
damain.farnell@manchester.ac.uk

Parkinson, J.B., Farnell, D.J.J.,: *An Introduction to Quantum Spin Systems*, Lect. Notes Phys. 816 (Springer, Berlin Heidelberg 2010), DOI 10.1007/978-3-642-13290-2

Lecture Notes in Physics ISSN 0075-8450   e-ISSN 1616-6361
ISBN 978-3-642-13289-6   e-ISBN 978-3-642-13290-2
DOI 10.1007/978-3-642-13290-2
Springer Heidelberg Dordrecht London New York

Library of Congress Control Number: 2010931613

Cover design: Integra Software Services Pvt. Ltd., Pondicherry

Printed on acid-free paper

Springer is part of Springer Science+Business Media (www.springer.com)

To Mary and Timothy;
and to Eve, Benjamin, and Lorna.

Preface

The topic of lattice quantum spin systems (or 'spin systems' for short) is a fascinating branch of theoretical physics and one of great pedigree, although many important questions still remain to be answered. The 'spins' are atomic-sized magnets that are localised to points on a lattice and they interact via the laws of quantum mechanics. This intrinsic quantum mechanical nature and the large (usually effectively infinite) number of spins leads to striking results which can be quite different from classical results and are often unexpected and indeed counter-intuitive.

Spin systems constitute the basic models of quantum magnetic insulators and so are relevant to a whole host of magnetic materials. Furthermore, they are important as prototypical models of quantum systems because they are conceptually simple and yet still demonstrate surprisingly rich physics. Low dimensional systems, in 2D and especially 1D, have been particularly fruitful because their simplicity has enabled exact solutions to be found which still contain many highly non-trivial features. Spin systems often demonstrate phase transitions and so we can use them to study the interplay of thermal and quantum fluctuations in driving such transitions. Of course there are many cases in which we can find no exact solution and in these cases they can be used as a testing ground for approximate methods of modern-day quantum mechanics. These quantum systems thus provide a great variety of interesting and difficult challenges to the mathematician or physical scientist.

This book was prompted by a series of talks given by one of the authors (JBP) at a summer school in Jyväskylä, Finland. These talks provided a detailed view of how one goes about solving the basic problems involved in treating and understanding spins systems at zero temperature. It was this level of detail, missing from other texts in the area, that prompted the other author (DJJF) to suggest that these lectures be brought together with supplementary material in order to provide a detailed guide which might be of use, perhaps to a graduate student starting work in this area.

The book is organised into chapters that deal firstly with the nature of quantum mechanical spins and their interactions. The following chapters then give a detailed guide to the solution of the Heisenberg and XY models at zero temperature using the Bethe Ansatz and the Jordan-Wigner transformation, respectively. Approximate methods are then considered from Chap. 7 onwards, dealing with spin-wave theory and numerical methods (such as exact diagonalisations and Monte Carlo). The coupled cluster method (CCM), a powerful technique that has only recently been applied to spin systems is described in some detail. The final chapter describes other work, some of it very recent, to show some of the directions in which study of these systems has developed.

The aim of the text is to provide a straightforward and practical account of all of the steps involved in applying many of the methods used for spins systems, especially where this relates to exact solutions for infinite numbers of spins at zero temperature. In this way, we hope to provide the reader with insight into the subtle nature of quantum spin problems.

Manchester, UK   John B. Parkinson
January 2010   Damian J.J. Farnell

Contents

Chapter 1
Introduction

Abstract This chapter introduces the concept of atomic magnets and how they combine to form systems which have macroscopic magnetic properties. Since the magnetic moment of a single atom is closely associated with its angular momentum or spin, it is necessary to study spins in order to understand macroscopic magnetism. At the atomic level, the properties of atoms are described in terms of quantum mechanics and hence we need to study quantum spins and their interactions (quantum spin systems). The nature of the exchange interaction is discussed, and we distinguish between ferromagnetic and antiferromagnetic coupling. Finally we summarise the contents of subsequent chapters.

Quantum spin systems are normally first encountered when studying magnetism. A magnet displays its familiar properties, for example attraction of metallic objects, because it is made up of atoms some or all of which are themselves tiny magnets. They produce the observed effects when all or most of these atomic magnets align in the same direction so that the net effect is the sum of all the individual magnets and is thus of macroscopic scale. In order for these atomic magnets to align there must be a force between them. One possibility for this force is the normal magnetic dipole interaction which can produce a tendency for two magnetic dipoles to align either parallel or antiparallel or indeed at an arbitrary angle, dependent upon their relative positions. This is a long range interaction extending over many atomic spacings. However it turns out that the magnetic dipole interaction is too weak to explain the phenomenon of room-temperature magnetism. If this were the only possible interaction, magnetism on a macroscopic scale would only occur at very low temperatures, within a few degrees of absolute zero, where thermal fluctuations are very weak and not strong enough to disrupt the delicate ordering.

In fact all magnetic materials that are of practical consequence have their atomic magnets aligned by means of the exchange interaction. This is a direct consequence of the Pauli exclusion principle and is basically due to the electric forces between electrons in the atoms rather than magnetic forces. Since forces of electrical origin are much stronger than those of magnetic origin at an atomic scale this explains why the exchange interaction is powerful enough to permit macroscopic scale magnetism even at room temperature. The exchange interaction has a much simpler form than the magnetic dipole interaction. In addition it is very short range, being dependent

Parkinson, J.B., Farnell, D.J.J.: *Introduction*. Lect. Notes Phys. **816**, 1–5 (2010)
DOI 10.1007/978-3-642-13290-2_1

on overlap of atomic wave functions. These two facts make it rather easier to handle than a magnetic dipole interaction.

The magnetic moment of an atom is often due, at least partly, to the motion of the electrons around the nucleus. In a classical picture we can imagine the electrons "orbiting" the nucleus rather like a planet or asteroid orbiting the sun. Since the electron has an electric charge this effectively creates a minute current loop around the nucleus and by the well known laws of electromagnetism this current loop has an associated magnetic dipole. All atoms except hydrogen have more than one electron and often the magnetic dipoles of the electrons cancel out so that most atoms do not possess a net magnetic dipole. Again thinking classically, the electron orbiting the nucleus also has an angular momentum, so we expect and find that the magnetic dipole of an atom, where it exists, is associated with the angular momentum of the atom. In fact the study of microscopic behaviour of magnetic materials is often a study of how the angular momenta of the atoms interact rather than the magnetic moments and that will be the case in this book.

Detailed calculation of the exchange interaction is possible in many cases. In practice the form can often be expressed as

$$\mathcal{E} = \alpha \mathbf{S}_1 . \mathbf{S}_2 + \beta S_1^z S_2^z \tag{1.1}$$

where $\mathcal{E}$ is the energy of interaction between two atoms with angular momentum $\mathbf{S}_1$ and $\mathbf{S}_2$. Here we use the symbol $\mathbf{S}$ for the vector angular momentum for reasons that will become clear shortly.

The special case $\alpha = J$, $\beta = 0$ is called the isotropic Heisenberg interaction. The special case $\alpha = 0$, $\beta = J$ is called the Ising interaction. In either case if J is negative the lowest energy is obtained when the angular momentum vectors are parallel. For the Heisenberg case there is no favoured axis of alignment, but for the Ising case the favoured alignment is along the z-axis. If J is positive then the lowest energy is obtained when the vectors are antiparallel rather than parallel.

Both special cases and the more general case have been extensively studied. Since the Ising interaction is simpler mathematically than the Heisenberg, results in 1D (i.e. one dimension) are usually fairly simple, but in 2D extremely important and non-trivial results have been obtained, in particular the Onsager solution [1]. Because the mathematical formalism used for this is very different to that used in this book it is not given here.

We use the word model to mean a system containing many atoms interacting with each other, usually by means of nearest-neighbour interaction only. Virtually no exact results are known for the 3D Ising model although there are extensive numerical studies. In the Heisenberg case, many important and again non-trivial exact results are known in 1D, whereas in 2D and 3D the results are almost entirely numerical.

When the interaction favours parallel alignment of the magnetic atoms the result is a ferromagnet. Antiparallel alignment gives rise to antiferromagnets, which do not have a net macroscopic magnetic moment. Nevertheless, at the microscopic level these materials are probably more interesting than ferromagnets and their study has contributed greatly to our understanding of magnetism, especially the quantum mechanical properties. Some properties of antiferromagnets do manifest themselves on a macroscopic scale, for example the magnetic contribution to the thermal properties such as the specific heat.

If the interaction is not restricted to nearest-neighbours, or there is more than one type of magnetic atom present in the material, then other types of magnetism can occur. A ferrimagnet is one in which some atoms are aligned in one direction and others in the opposite direction, but there is a mismatch in the total magnetic moment in the two directions so that there is a net macroscopic moment. Spiral ordering can also occur in which the atoms do not align all along one axis, but the direction rotates as one moves through the material. Often, but not always, this results in a non-zero macroscopic magnetic moment. Ferrimagnets and spiral magnets will not be covered in this book.

In this book we shall consider spin systems. These are systems that consist of two or more particles each of which has an angular momentum and associated magnetic moment. As a result, study of these systems is effectively a study of certain types of magnetic materials. Usually the magnetic particles are atoms or ions of transition metals such as Mn, Ni or Fe or rare-earths such as Ce or Nd. The book by Mattis [2] has an excellent account of the history of magnetism.

Because this is such a vast topic we immediately restrict our attention to insulating materials, i.e. we shall not consider metallic magnets. We take the particles to be interacting by Heisenberg exchange which is electrical in origin and is normally much stronger than a magnetic interaction. The exchange interaction is in fact between the angular momenta of the particles but since their magnetic moment is proportional to the angular momentum it superficially appears as an interaction between the magnetic moments, albeit of a much simpler form than the magnetic dipole interaction.

We shall mainly consider the quantum case although there are many interesting classical results which we shall include in some places.

Heisenberg in 1925 [3] and Schrödinger in 1926 [4–6] established the basis of modern quantum mechanics. Their formalisms are somewhat different although equivalent. The Schrödinger formalism emphasises the explicit form of the wave function as obtained from the Schrödinger equation whereas the Heisenberg formalism expresses the wave function as a vector which represents a linear combination of basis states. The former is very successful in particular in atomic physics and in problems in which particles move in a potential well with or without the presence of electric and magnetic fields.

One area in which the Heisenberg formalism is especially useful is in the description of angular momentum. For the orbital angular momentum of single particles this can be done in the Schrödinger formalism, but for interacting particles, especially many-body systems, it is much more convenient to use the Heisenberg formalism. Indeed, for the intrinsic angular momentum of elementary particles, the spin, it is not possible to obtain a specific wave function. For baryons such as the neutron and proton this is because the internal structure in terms of quarks is rather complicated and in any case the quarks themselves have a spin, while the leptons such as the electron are believed to be point particles with no internal structure.

Most of the magnetism observed in practice is due to electrons. It may be associated with either the orbital angular momentum or the spin. For simplicity we shall refer to the angular momenta in all cases as spin even when it is wholly or partly due to orbital motion. The constant of proportionality between the magnetic moment and the angular momentum is different in the two cases and we write

$$\boldsymbol{\mu} = g\beta\mathbf{S}$$

where $\boldsymbol{\mu}$ is the magnetic moment, $\mathbf{S}$ is the spin measured in units of $\hbar$, $\beta = \dfrac{e\hbar}{2m}$ is the elementary unit of magnetic moment associated with unit spin, the Bohr magneton, and g is the gyromagnetic ratio which is 1 for orbital angular momentum, 2 for intrinsic spin and has an effective intermediate value when both are present.

We shall deal mainly with the ground state of the system and the elementary excitations from this. This implies $T = 0$ or $T \to 0$ and does not allow non-zero T for which much more detailed knowledge of the full eigenstates is needed. The exception to this is the 1D chain with $S = \frac{1}{2}$ where the extra information needed is available through the Bethe ansatz.

We shall consider all dimensionalities: 0D, 1D, 2D and 3D. 0D refers to finite-sized clusters or chains for which exact results can often be obtained. These are becoming of more interest in their own right but they can also shed light on the behaviour in higher dimensionalities. 1D (i.e. infinite chains) is the dimension in which dramatic exact quantum results for $S = \frac{1}{2}$ have been obtained.

Numerical results include the Density Matrix Renormalisation Group (DMRG) [7], which has been superbly successful in 1D where exact results are not available, e.g. for $S > \frac{1}{2}$. Many other numerical techniques exist, which are particularly important for 2D and 3D where it has proved difficult to use the DMRG. We shall describe one of these, the coupled cluster method, in detail in Chap. 10.

We shall mainly consider antiferromagnetic coupling with exchange interaction $J > 0$. For ferromagnetic coupling with negative J it is usually straightforward to write down the ground state in both quantum and classical cases. The elementary excitations are also usually simple to obtain, often in the form of spin waves or bound pairs of spin waves. The antiferromagnetic problem is much more difficult and also more interesting and shows much greater differences between the classical and quantum systems as is known from the analysis of the $S = \frac{1}{2}$ chain. We shall restrict our discussion mainly to systems with only nearest-neighbour exchange, although next-nearest-neighbour exchange is considered briefly in the final chapter. However, we shall allow the exchange to be anisotropic, and we shall also consider the effect of an applied magnetic field in some places.

Another interesting feature of antiferromagnetic systems is the possibility of frustration. This occurs when the system contains nearest-neighbour paths with odd numbers of atoms. The simplest example is a cluster of three atoms, each of which is nearest neighbour to both of the others. Since the antiferromagnetic exchange favours antiparallel alignment of nearest neighbours and this is not possible for all neighbours we use the term frustration. This can also happen in larger finite-sized clusters such as a five atom ring, a tetrahedron or an octahedron. It does not happen in 1D for nearest-neighbour-only exchange but it occurs in 2D, for example in triangular and Kagome lattices and in 3D, for example in the HCP lattice.

Again we emphasise that this is not intended to be a comprehensive account of spin systems which would be a formidable undertaking. Rather it is intended to fulfil two roles. Firstly we present an introduction to spin systems which should be of use to newcomers to the area, particularly graduate students. We then give a fairly detailed description of the $S = \frac{1}{2}$ chain which, because it has yielded so many exact results of a highly non-trivial and counter-intuitive nature, can reasonably be described as one of the crowning glories of many-body physics. Secondly we consider some of the most important approximate methods including an important modern technique, the coupled cluster method, and finally look at other work on these systems.

Lastly, we mention that there are, of course, numerous other books on magnetism and spin systems [2, 8–14]. As well as the the very comprehensive account given in Mattis [2], an excellent introductory text is the book by Caspers [8]. The present book differs from Caspers, partly in the choice of topics but also in the attempt to give complete details of the mathematics of some of the important results.

References

1. Onsager, L.: Phys. Rev. **65**(2), 117–149 (1944)
2. Mattis, D.C.: The Theory of Magnetism I, Springer, Berlin (1981, 1988)
3. Heisenberg, W.Z.: Physik **33**, 879 (1925)
4. Schrödinger, E.: Ann. Phys. **79**, 361, 489, 784 (1926)
5. Schrödinger, E.: Ann. Phys. **80**, 437 (1926)
6. Schrödinger, E.: Ann. Phys. **81**, 109 (1926)
7. White, S.R.: Phys. Rev. Lett. **69**, 2863 (1992)
8. Caspers, W.J.: Spin Systems. World Scientific, Singapore (1989)
9. Schollwöck, U., Richter, J., Farnell, D.J.J., Bishop, R.F. (eds.): Quantum Magnetism. Lecture Notes in Physics, vol. 645. Springer, Berlin (2004)
10. Diep, T.H.: Frustrated Spin Systems. World Scientific, Singapore (2004)
11. Auerbach A.: Interacting Electrons and Quantum Magnetism. Springer, New York, NY (1994)
12. Majlis, N.: The Quantum Theory of Magnetism. World Scientific, Singapore (2007)
13. Läuchli, A.M.: Quantum Magnetism and Strongly Correlated Electrons in Low Dimensions. Swiss Federal Institute of Technology, Zurich (2002)
14. White, R.M.: Quantum Theory of Magnetism. Springer, Berlin (1983)

Chapter 2
Spin Models

Abstract Assuming only a basic knowledge of quantum mechanics, we develop the quantum mechanics of angular momentum, distinguishing between orbital and intrinsic angular momentum or spin. Suitable basis states are introduced, as are the fundamental operators which act upon these. We consider the situation of just two spins interacting via an exchange interaction. This leads us to introduce other spin operators and evaluate their commutators. We discuss what happens when there are large numbers of interacting spins. Finally we introduce the infinite linear (i.e. one-dimensional) chain of spin-1/2 atoms. Some preliminary classical results for this model are given as background for the full quantum treatment which will be studied in detail in the following chapters.

2.1 Spin Angular Momentum

In this section we give a brief summary of some of the important results about angular momentum [1] which are needed for a study of quantum spin systems.

There are two types of angular momentum which occur in nature. The first of these, in its classical form, is familiar from very early days and arises from motion, often circular, relative to some axis. This is called *orbital* angular momentum.

In quantum mechanics we usually first come across orbital angular momentum when we study the eigenstates of the Schrödinger equation (SE) for the hydrogen atom. The angular parts of these, using spherical polar coordinates, have the form

$$Y_{\ell,m}(\theta, \phi) = C_{\ell,m} P_\ell^m(\theta) e^{im\phi} ,$$

where P_ℓ^m are polynomials in $\cos\theta$ and/or $\sin\theta$ of degree ℓ – the associated Legendre polynomials and $C_{\ell,m}$ is a normalisation constant.

These angular parts of the eigenstates of the time-independent SE have another significance. They are also eigenstates of the angular momentum operators. For a particle with momentum $\mathbf{p}$ and position vector $\mathbf{r}$ then the classical angular momentum relative to an axis through the origin and perpendicular to the plane of motion is $\mathbf{L} = \mathbf{r} \times \mathbf{p}$, and in quantum mechanics the angular momentum operator has the same form

Parkinson, J.B., Farnell, D.J.J.: *Spin Models*. Lect. Notes Phys. **816**, 7–19 (2010)
DOI 10.1007/978-3-642-13290-2_2 © Springer-Verlag Berlin Heidelberg 2010

$$\hat{\mathbf{L}} = \hat{\mathbf{r}} \times \hat{\mathbf{p}}$$

where $\hat{}$ denotes an operator. These operators have the commutators $\left[\hat{L}^x, \hat{L}^y\right] = i\hbar\hat{L}^z$ together with cyclic permutations.

Two of the most useful operators are $\hat{L}^z$ and $\hat{\mathbf{L}}^2 = (\hat{L}^x)^2 + (\hat{L}^y)^2 + (\hat{L}^z)^2$. This is because the $Y_{\ell,m}$ are eigenstates of both of these simultaneously. It is always possible to find states which are simultaneous eigenstates of two operators provided the operators commute and we can easily show that $[\hat{L}^z, \hat{\mathbf{L}}^2] = 0$. The corresponding eigenvalues are given by

$$\hat{L}^z Y_{\ell,m} = m\hbar Y_{\ell,m}; \qquad \hat{\mathbf{L}}^2 Y_{\ell,m} = \ell(\ell+1)\hbar^2 Y_{\ell,m}.$$

There are two other operators which have useful properties. These are defined as

$$\hat{L}^+ \equiv \hat{L}^x + i\hat{L}^y \quad \text{and} \quad \hat{L}^- \equiv \hat{L}^x - iL^y.$$

From these definitions it follows that

$$[\hat{L}^-, \hat{L}^+] = 2\hbar L^z$$

and it can also be shown that

$$\hat{L}^+ Y_{\ell,m} = 0 \quad \text{if } m = \ell.$$

However, if $m < \ell$ then

$$\hat{L}^z(\hat{L}^+ Y_{\ell,m}) = (m+1)\hbar(L^+ Y_{\ell,m}),$$
$$\hat{\mathbf{L}}^2(\hat{L}^+ Y_{\ell,m}) = \ell(\ell+1)\hbar^2(\hat{L}^+ Y_{\ell,m}).$$

This means that $L^+ Y_{\ell,m}$ is also an eigenstate of $\hat{L}^z$ and $\hat{\mathbf{L}}^2$. It has the same eigenvalue of $\hat{\mathbf{L}}^2$ as $Y_{\ell,m}$, but an eigenvalue of $\hat{L}^z$ which is increased by one unit of $\hbar$. Clearly $\hat{L}^+$ converts $Y_{\ell,m}$ into $Y_{\ell,m+1}$, to within a multiplicative constant, and it is called the *raising* operator. Similarly $\hat{L}^-$ converts $Y_{\ell,m}$ into $Y_{\ell,m-1}$ to within a multiplicative constant and is a *lowering* operator. That is $\hat{L}^+ Y_{\ell,m} = A Y_{\ell,m+1}$ and $\hat{L}^- Y_{\ell,m} = B Y_{\ell,m-1}$.

The second type sort of angular momentum which occurs in nature is the intrinsic spin of elementary particles. This is a fundamental property of a particle. It is often thought of as due to internal spinning of a particle, hence the name, but this is not correct since, for example, an electron has spin angular momentum but is believed to be a point particle with no internal structure.

The spin angular momentum has most of the properties of orbital angular momentum, but also some special properties. For the present we shall use $\hat{\mathbf{S}}$ for spin angular momentum instead of $\hat{\mathbf{L}}$. Again we can use the eigenstates of $\hat{S}^z$ and $\hat{S}^2$ with the same quantum numbers ℓ and m such that

$$\hat{S}^z|\ell, m\rangle = m\hbar|\ell, m\rangle$$
$$\hat{\mathbf{S}}^2|\ell, m\rangle = \ell(\ell + 1)\hbar^2|\ell, m\rangle$$

Clearly the spin eigenstate $|\ell, m\rangle$ behaves here just like the orbital eigenstate $Y_{\ell, m}$, but there are two fundamental differences.

1. $|\ell, m\rangle$ does not have an explicit form in terms of θ and ϕ like $Y_{\ell, m}$.
2. ℓ may now be an integer, as before, *or* an integer $+\frac{1}{2}$, sometimes called a 'half-integer'.

The values of m still differ by unity and still take all values between $-\ell$ and $+\ell$ but if ℓ is an integer $+\frac{1}{2}$ then so are the m.

For example 1 If $\ell = \frac{3}{2}$ then the possible values of m are $-\frac{3}{2}$, $-\frac{1}{2}$, $\frac{1}{2}$, and $\frac{3}{2}$.

For example 2 If $\ell = \frac{1}{2}$ then m can be $-\frac{1}{2}$ or $\frac{1}{2}$.

We also have the operators $\hat{S}^+$ and $\hat{S}^-$ corresponding to $\hat{L}^+$ and $\hat{L}^-$ and with the same properties as before, namely

$$\hat{S}^+|\ell, m\rangle = 0 \quad \text{for} \quad m = \ell$$

while for $m < \ell$

$$\hat{S}^z(\hat{S}^+|\ell, m\rangle) = (m + 1)\hbar(\hat{S}^+|\ell, m\rangle)$$
$$\hat{\mathbf{S}}^2(\hat{S}^+|\ell, m\rangle) = \ell(\ell + 1)\hbar^2(S^+|\ell, m\rangle).$$

Again there are similar results for $\hat{S}^-|\ell, m\rangle$.

For the orbital angular momentum of an electron in a H-atom, states with different ℓ are possible, each of which has several possible values of m. For spin angular momentum the value of ℓ is *fixed* (it is a fundamental property of the particle) e.g. $\ell = \frac{1}{2}$ for electron $\ell = 1$ for some mesons etc. The only difference between eigenstates is the value of m.

In magnetic materials the magnetic moment of an atom may be due to a combination of spin and orbital angular momentum. The actual magnetic moment is a simple but non-trivial function of the two types, usually described in terms of the 'Landé g-factor' of the atom or ion which varies between 1 (for pure orbital angular momentum) and 2 (for pure spin angular momentum). However we shall always take the angular momentum to have a fixed value of ℓ (i.e. the 'magnitude') but allow m (i.e. the 'orientation') to vary. For this reason we usually refer to the angular momentum of a magnetic atom as a spin and use the symbol $\hat{\mathbf{S}}$, even when it is a combination of both types.

Single Spin-$\frac{1}{2}$

For a single spin-$\frac{1}{2}$ the value of ℓ is $\frac{1}{2}$. The value of m may be $+\frac{1}{2}$ or $-\frac{1}{2}$, so there are two eigenstates of $\hat{S}^z$:

$$|\ell, m\rangle = |\tfrac{1}{2}, \tfrac{1}{2}\rangle \quad \text{or} \quad |\tfrac{1}{2}, -\tfrac{1}{2}\rangle.$$

Alternative abbreviated notations for these which we use are

$$|\tfrac{1}{2}, \tfrac{1}{2}\rangle \equiv |\tfrac{1}{2}\rangle \equiv |+\rangle$$
$$|\tfrac{1}{2}, -\tfrac{1}{2}\rangle \equiv |-\tfrac{1}{2}\rangle \equiv |-\rangle.$$

These two eigenstates form a complete set for a single spin $\tfrac{1}{2}$ so an arbitrary state may be written

$$\psi = \alpha|+\rangle + \beta|-\rangle.$$

The eigenstates are normalized and orthogonal:

$$\langle +|+\rangle = \langle -|-\rangle = 1$$
$$\langle +|-\rangle = 0.$$

From now on we shall omit the $\hat{\ }$ symbol on operators and also work in units in which $\hbar$ has the value 1. Hence

$$S^z|+\rangle = \tfrac{1}{2}\hbar|+\rangle = \tfrac{1}{2}|+\rangle$$
$$S^z|-\rangle = -\tfrac{1}{2}\hbar|-\rangle = -\tfrac{1}{2}|+\rangle$$

Also

$$S^+|+\rangle = 0 \qquad S^-|-\rangle = 0$$
$$S^+|-\rangle = |+\rangle \qquad S^-|+\rangle = |-\rangle$$

Single Spin-1 This follows a similar pattern with $\ell = 1$ and $m = 1, 0, -1$. Again we use a simplified notation:

$$|1, 1\rangle \equiv |1\rangle \equiv |+\rangle$$
$$|1, 0\rangle \equiv |0\rangle$$
$$|1, -1\rangle \equiv |-1\rangle \equiv |-\rangle$$

and the effect of the spin operators is

$$S^+|-\rangle = \sqrt{2}|0\rangle \quad S^+|0\rangle = \sqrt{2}|+\rangle \quad S^+|+\rangle = 0$$
$$S^-|-\rangle = 0 \qquad S^-|0\rangle = \sqrt{2}|-\rangle \quad S^-|+\rangle = \sqrt{2}|0\rangle$$
$$S^z|-\rangle = -|-\rangle \quad S^z|0\rangle = 0 \qquad S^z|+\rangle = |+\rangle.$$

There are similar results for higher spins.

2.2 Coupled Spins

When spins interact with each other new phenomena can arise such as large scale ferromagnetism or antiferromagnetism. If we study this behaviour using quantum mechanics then some results are very similar to what one would expect from a classical point of view. However, some results are quite different and this is what makes the study of interacting spin systems so fascinating.

The quantum treatment involves quite a lot of mathematics. In fact advanced treatments use some very sophisticated methods, but in this introductory text these will be kept to a minimum.

Much of our understanding comes from a study of exactly solvable models. These are rather specialised but because they can be solved exactly they can be studied in great depth. In 1D we shall consider in later chapters two models in particular:

1. the Heisenberg chain – studied using the 'Bethe Ansatz'   and
2. the XY chain – studied using the 'Jordan-Wigner transformation'.

In the rest of this chapter we look at what happens when small numbers of spins interact with each other, in particular the similarities and differences between the classical and quantum mechanical cases. Simple results for much larger systems will be introduced, setting the stage for the more complicated infinite quantum systems discussed in later chapters.

2.3 Two Interacting Spin-$\frac{1}{2}$'s

Suppose we have just two atoms, each with spin-$\frac{1}{2}$, interacting with an isotropic exchange interaction J (Heisenberg Exchange). The Hamiltonian is

$$\mathcal{H} = J\,\mathbf{S}_1 \cdot \mathbf{S}_2 \tag{2.1}$$

Classically the energy of the system depends on the angle between the spins since

$$\mathbf{S}_1 \cdot \mathbf{S}_2 = S_1 S_2 \cos\theta.$$

S_1 and S_2 are both $\frac{1}{2}$ so the energy is

$$E_{\text{class}} = \frac{1}{4}\,J\,\cos\theta,$$

and since $-1 \leq \cos\theta \leq 1$ all energies are possible between $+\frac{1}{4}J$ and $-\frac{1}{4}J$.

Quantum mechanically we need to find all the energy eigenstates of the system. First we write down a complete orthonormal set of states for the system. Then we write $\mathcal{H}$ as a matrix using this basis and finally diagonalise to find the eigenstates.

Using subscript 1 for the first spin, then a basis for the first spin consists of the two eigenstates of S_1^z, namely $|+\rangle_1$ and $|-\rangle_1$. Similarly $|+\rangle_2$ and $|-\rangle_2$ form a basis for the second spin. A basis for the pair is then

$$\{|++\rangle,\ |+-\rangle,\ |-+\rangle,\ |--\rangle\} \tag{2.2}$$

where $|++\rangle$ means $|+\rangle_1\,|+\rangle_2$ etc. Hence the matrix for $\mathcal{H}$ will be size 4×4.

It is useful to rewrite the Hamiltonian (2.1) as follows

$$\mathcal{H} = J\,\mathbf{S}_1 \cdot \mathbf{S}_2$$
$$= J\,(S_1^z S_2^z + S_1^x S_2^x + S_1^y S_2^y)$$
$$= J\left(S_1^z S_2^z + \frac{1}{2}S_1^+ S_2^- + \frac{1}{2}S_1^- S_2^+\right) \qquad (2.3)$$

This is easily shown as follows:

$$S_1^+ S_2^- = (S_1^x + i S_1^y)(S_2^x - i S_2^y)$$
$$= S_1^x S_2^x + S_1^y S_2^y + i(S_1^y S_2^x - S_1^x S_2^y)$$
$$\text{and } S_1^- S_2^+ = S_1^x S_2^x + S_1^y S_2^y - i(S_1^y S_2^x - S_1^x S_2^y)$$
$$\text{so } S_1^+ S_2^- + S_1^- S_2^+ = 2(S_1^x S_2^x + S_1^y S_2^y).$$

Now consider the effect of the Hamiltonian (2.3) operating on each member of the basis (2.2), noting the following points

1. All operators on *different* atoms commute e.g. $[S_1^x, S_2^y] = 0$.
2. S_1 operators act only on the first spin, and have no effect on the second

$$e.g. \quad S_1^+|--\rangle = |+-\rangle,$$

and likewise S_2 operators act only on the second spin.

First consider $S_1^z S_2^z$ operating on each basis state:

$$S_1^z S_2^z|++\rangle = \frac{1}{2}\cdot\frac{1}{2}|++\rangle \quad = \quad \frac{1}{4}|++\rangle$$
$$S_1^z S_2^z|+-\rangle = -\frac{1}{4}|+-\rangle$$
$$S_1^z S_2^z|-+\rangle = -\frac{1}{4}|-+\rangle$$
$$S_1^z S_2^z|--\rangle = +\frac{1}{4}|--\rangle.$$

Now consider $S_1^+ S_2^-$ and $S_1^- S_2^+$ operating on each basis state, noting that $S_1^+|++\rangle = 0$ since S_1^+ acts on $|+\rangle_1$ etc.:

$$S_1^+ S_2^-|+-\rangle = 0$$
$$S_1^+ S_2^-|-+\rangle = |+-\rangle$$
$$S_1^+ S_2^-|--\rangle = 0$$
$$S_1^- S_2^+|++\rangle = 0$$
$$S_1^- S_2^+|+-\rangle = |-+\rangle$$
$$S_1^- S_2^+|-+\rangle = S_1^- S_2^+|--\rangle = 0$$

Using these together with (2.3) we get

$$\mathcal{H}|++\rangle = \frac{1}{4}J|++\rangle$$

$$\mathcal{H}|+-\rangle = -\frac{1}{4}J|+-\rangle + \frac{1}{2}J|-+\rangle$$

$$\mathcal{H}|-+\rangle = -\frac{1}{4}J|-+\rangle + \frac{1}{2}J|+-\rangle$$

$$\mathcal{H}|--\rangle = \frac{1}{4}J|--\rangle,$$

which means that in matrix form, labelling the rows with the basis states

$$\mathcal{H} = J \begin{pmatrix} \frac{1}{4} & 0 & 0 & 0 \\ 0 & -\frac{1}{4} & \frac{1}{2} & 0 \\ 0 & \frac{1}{2} & -\frac{1}{4} & 0 \\ 0 & 0 & 0 & \frac{1}{4} \end{pmatrix} \begin{array}{l} |++\rangle \\ |+-\rangle \\ |-+\rangle \\ |--\rangle \end{array}$$

The eigenvalues of this matrix are easily obtained (omitting J for clarity in the working)

$$\begin{vmatrix} \frac{1}{4} - E & 0 & 0 & 0 \\ 0 & -\frac{1}{4} - E & \frac{1}{2} & 0 \\ 0 & \frac{1}{2} & -\frac{1}{4} - E & 0 \\ 0 & 0 & 0 & \frac{1}{4} - E \end{vmatrix} = 0$$

$$\left(\frac{1}{4} - E\right)^2 \begin{vmatrix} -\frac{1}{4} - E & \frac{1}{2} \\ \frac{1}{2} & -\frac{1}{4} - E \end{vmatrix} = 0$$

$$\left(\frac{1}{4} - E\right)^2 \left[\left(\frac{1}{4} + E\right)^2 - \frac{1}{4}\right] = 0$$

$$\left(\frac{1}{4} - E\right)^2 \left[\left(\frac{1}{4} + E\right) + \frac{1}{2}\right]\left[\left(\frac{1}{4} + E\right) - \frac{1}{2}\right] = 0$$

giving $E = \frac{1}{4}, \frac{1}{4}, \frac{1}{4}, -\frac{3}{4}$.

Thus, after restoring J, the eigenvalues are $E = \dfrac{J}{4}$ (3 times) and $-\dfrac{3J}{4}$. The first three of these form a triplet and the last one a singlet.

The maximum eigenvalue here, $\frac{J}{4}$, is the same as in the classical case, but the minimum eigenvalue, $-\frac{3J}{4}$, is much lower. This is clearly a new quantum effect.

Now let us consider the corresponding eigenstates. Since the three eigenvalues $\left(\frac{J}{4}\right)$ are degenerate the choice of corresponding eigenvectors is somewhat arbitrary. It is easy to verify that the following is a suitable orthonormal set.

$$\psi_1 \equiv \begin{pmatrix} 1 \\ 0 \\ 0 \\ 0 \end{pmatrix} \qquad \psi_2 \equiv \begin{pmatrix} 0 \\ \frac{1}{\sqrt{2}} \\ \frac{1}{\sqrt{2}} \\ 0 \end{pmatrix} \qquad \psi_3 \equiv \begin{pmatrix} 0 \\ 0 \\ 0 \\ 1 \end{pmatrix} \qquad \psi_4 \equiv \begin{pmatrix} 0 \\ \frac{1}{\sqrt{2}} \\ -\frac{1}{\sqrt{2}} \\ 0 \end{pmatrix}$$

the first three having eigenvalue $\frac{J}{4}$ and ψ_4 having eigenvalue $-\frac{3J}{4}$.

2.4 Commutators and Quantum Numbers

The commutation relations between the operators S^z, S^+ and S^- follow from those between S^x, S^y and S^z. From the definitions of $S^+ \equiv S^x + iS^y$ and $S^- \equiv S^x - iS^y$ we have

$$[S^z, S^+] = [S^z, S^x] + i[S^z, S^y]$$
$$= i\hbar S^y + i(-i\hbar S^x)$$
$$= \hbar S^+ = S^+ \text{ (putting } \hbar = 1 \text{ again).}$$

Similarly $[S^z, S^-] = -S^-$,
and $[S^-, S^+] = -i[S^y, S^x] + i[S^x, S^y] = -2S^z$.

For the interacting pair of spins the operator

$$S_T^z = S_1^z + S_2^z$$

gives the z-component of total angular momentum of the pair.

Clearly $[S_T^z, S_1^z S_2^z] = 0$ since only S_1^z and S_2^z are involved. Also

$$[S_T^z, S_1^+ S_2^-] = [S_1^z, S_1^+ S_2^-] + [S_2^z, S_1^+ S_2^-]$$
$$= [S_1^z, S_1^+]S_2^- + S_1^+[S_2^z, S_2^-]$$
$$= S_1^+ S_2^- + S_1^+(-S_2^-) \quad = \quad 0,$$

and similarly $[S_T^z, S_1^- S_2^+] = 0$.

This means that S_T^z commutes with *each* term in Hamiltonian (2.3), so S_T^z is a good quantum number. In fact the basis states have the property

$$S_T^z |++\rangle = \left(\frac{1}{2} + \frac{1}{2}\right)|++\rangle = |++\rangle$$

$$S_T^z |+-\rangle = \left(\frac{1}{2} - \frac{1}{2}\right)|+-\rangle = 0$$

$$S_T^z |-+\rangle = 0$$

$$S_T^z |--\rangle = -|--\rangle$$

so the eigenstates ψ_1, ψ_2, ψ_3 and ψ_4 have the properties that

$$S_T^z \psi_1 = \psi_1, \quad S_T^z \psi_2 = 0, \quad S_T^z \psi_3 = -\psi_3 \quad \text{and} \quad S_T^z \psi_4 = 0.$$

ψ_2 and ψ_4 have eigenvalue 0 of S_T^z while ψ_1 has eigenvalue 1 and ψ_3 eigenvalue -1.

For Heisenberg exchange we also have one additional property. The Hamiltonian (2.3) is

$$\mathcal{H} = J\mathbf{S}_1 \cdot \mathbf{S}_2$$

which is obviously isotropic and does not distinguish between x, y, z. Since we know that $[S_T^z, \mathcal{H}] = 0$ it follows that $[S_T^x, \mathcal{H}] = 0$ and $[S_T^y, \mathcal{H}] = 0$ also.

If two operators commute it is possible to find states which are eigenstates of both simultaneously. Since $[S_T^x, S_T^z] \neq 0$ we can find states which are eigenstates of $\mathcal{H}$ and also of S_T^z as we have done and we could find (different) states which are eigenstates of $\mathcal{H}$ and also of S_T^x. However it is not possible to find states which are eigenstates of both S_T^x and S_T^z.

The total spin angular momentum is given by

$$\mathbf{S}_T = \mathbf{S}_1 + \mathbf{S}_2$$
$$\text{so} \quad \mathbf{S}_T^2 = (\mathbf{S}_1 + \mathbf{S}_2)^2 = S_T^{x^2} + S_T^{y^2} + S_T^{z^2}.$$

Since $\mathcal{H}$ commutes with all of S_T^x, S_T^y and S_T^z it follows that $[\mathbf{S}_T^2, \mathcal{H}] = 0$ also. The eigenstates $\{\psi_1, \psi_2, \psi_3, \psi_4\}$ are simultaneously eigenstates of $\mathcal{H}$, S_T^z and $\mathbf{S}_T^2$.

It may be easily verified that the corresponding eigenvalues of the operator $\mathbf{S}_T^2$ are given by

$$\mathbf{S}_T^2 \psi_1 = 2\psi_1$$
$$\mathbf{S}_T^2 \psi_2 = 2\psi_2$$
$$\mathbf{S}_T^2 \psi_3 = 2\psi_3$$
$$\mathbf{S}_T^2 \psi_4 = 0.$$

Hence the three triplet states ψ_1, ψ_2, ψ_3 have total angular momentum quantum number $\ell = 1$, i.e. $\mathbf{S}_T^2 \psi_1 = \ell(\ell+1)\psi_1$ with $\ell = 1$, while the singlet state ψ_4 has $\ell = 0$.

2.5 Physical Picture

A simple physical picture of this situation is as follows. Combining two spin-$\frac{1}{2}$ atoms, we can form either a 'spin-1' configuration or a 'spin-0' configuration. The 'spin-1' configuration can have 3 possible 'orientations', i.e. values of z component, giving

state	explicit form	'picture'	S_T	S_T^z	E	
ψ_1	$\lvert ++\rangle$	↑↑	1	1	$J/4$	
ψ_2	$\frac{1}{\sqrt{2}}(\lvert +-\rangle + \lvert -+\rangle)$	→→	1	0	$J/4$	triplet
ψ_3	$\lvert --\rangle$	↓↓	1	−1	$J/4$	
ψ_4	$\frac{1}{\sqrt{2}}(\lvert +-\rangle - \lvert -+\rangle)$	↑↓	0	0	$-3J/4$	singlet

The three triplet states have the two spins parallel. This is called a *ferromagnetic* arrangement since the magnetic moments will also be parallel.

The singlet ψ_4 is a 'spin-0' configuration which has no 'orientation'. This state has antiparallel spins and is an *antiferromagnetic* arrangement.

It follows that if J is positive then the ground state is antiferromagnetic, but if J is negative then the ground state is ferromagnetic. This agrees with the classical picture, but the actual values of the energy are not the same as in the classical case.

2.6 Infinite Arrays of Spins

A magnetic crystal consists of a large number, N, of magnetic atoms in a regular array. Each atom has a 'spin' and associated magnetic moment. If the spins (and moments) are all aligned in the same direction we have a *ferromagnet* with a large net magnetic moment. Other possible configurations in the absence of a magnetic field are

$$\text{disordered (random alignment)} \longrightarrow \text{paramagnet}$$
$$\text{alternating up and down} \longrightarrow \text{antiferromagnet}$$

If a small magnetic field is applied to a paramagnet then the ordering will still be largely random but with a tendency to align in the opposite direction to the applied field.

The most important case in practice is the limit $N \to \infty$, called the *thermodynamic limit*. At any non-zero temperature, thermal fluctuations will tend to reduce the perfect alignment in a ferromagnet or antiferromagnet. However, there may still be a net alignment up to some critical temperature T_C above which the system is paramagnetic. This critical temperature is called the Curie temperature for a ferromagnet, and the Néel temperature, usually written as T_N, for an antiferromagnet (Fig. 2.1).

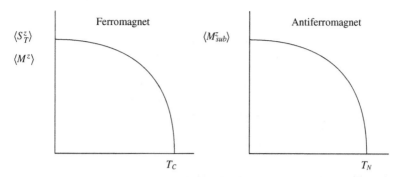

Fig. 2.1 Magnetisation of a ferromagnet and sublattice magnetisation of an antiferromagnet as a function of temperature. Note that the total magnetisation $\langle M^z \rangle$ is always zero for the antiferromagnet

In one dimension (1D) and two dimensions (2D) the critical temperature T_C or T_N is zero, i.e. $\langle M^z \rangle = 0$ or $\langle M^z_{\text{sub}} \rangle = 0$ for $T > 0$, where $M^z = g \beta S^z_T$ is the magnetisation, but at $T = 0$ we may still have ferromagnetic or antiferromagnetic ordering. In addition in 2D there occurs a more subtle type of ordering at non-zero temperatures known as the Kosterlitz–Thouless transition [2].

As we discussed earlier, the reason why atoms try to align parallel (or anti-parallel) is because of the interaction between them. Although this could be a magnetic dipole interaction, in practice it is usually an exchange interaction which is quantum mechanical in origin and derives from the electric Coulomb force between electrons and hence is much stronger than the magnetic dipole interaction.

Depending on the types of atom involved and the environment in which they exist the exchange interaction may have different forms. Examples are:

a. Heisenberg $J\, \mathbf{S}_1 \cdot \mathbf{S}_2$ (as before)
b. Ising $J S^z_1 S^z_2$
c. Anisotropic (a combination of the above)

$$J[\Delta S^z_1 S^z_2 + (S^x_1 S^x_2 + S^y_1 S^y_2)]$$

d. Biquadratic $J(\mathbf{S}_1 \cdot \mathbf{S}_2)^2$

In addition there may be other terms in the Hamiltonian which are not interactions, but involve individual spins e.g. crystal field terms, typically of the form $A(S^z_1)^2$, or external magnetic field terms of the form $H S^z_1$.

A magnetic crystal consists of a large array with many (N) atoms. However, the exchange interaction is normally very short range. A simple approximation, but one that is accurate for many crystals, is to assume that only nearest-neighbours in the array interact. The simplest types of array are

3D Simple cubic. A interacts with B, C and D but not with E or F.

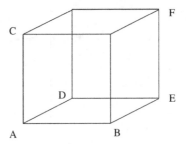

2D Square. A interacts with B but not with C.

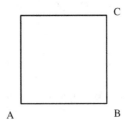

1D Chain. A interacts with B but not with C.

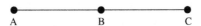

The differences between quantum mechanical and classical behaviour are most marked for low values of spin and for low dimensions. For these reasons, and also because it is the most tractable mathematically, we shall study first the 1D chain of spin-$\frac{1}{2}$ atoms interacting with nearest neighbour Heisenberg exchange.

2.7 1D Heisenberg Chain with $S = \frac{1}{2}$ and Nearest-Neighbour Interaction

Since this is a 1D system the ferromagnetic or antiferromagnetic ordering only occurs at zero temperature but as well as being simpler to handle than non-zero T it is arguably more interesting.

The details of how this chain is treated quantum mechanically is the subject of the next chapters. Here we simply set the stage by giving some simple classical results.

The mathematics is simplified if we use periodic boundary conditions. This means that the ends of the chain are joined so that the N'th atom is a nearest-neighbour of the first as well as of the $(N-1)$'th. Of course the chain is not actually curved to achieve this: it is purely a mathematical device which ensures that there are no 'ends' to worry about. In any case, we shall take the limit $N \to \infty$, so any curvature would tend to zero.

The Hamiltonian for the system is

$$\mathcal{H} = J[\mathbf{S}_1 \cdot \mathbf{S}_2 + \mathbf{S}_2 \cdot \mathbf{S}_3 + \ldots + \mathbf{S}_N \cdot \mathbf{S}_1]$$

$$= J \sum_{i=1}^{N} \mathbf{S}_i \cdot \mathbf{S}_{i+1} \quad \text{where } i + N \equiv i \tag{2.4}$$

Since the interaction is nearest-neighbour only, the classical solution of (2.4) is very similar to the case of just two spins.

If $J < 0$ the lowest energy of a pair of spins occurs when they are aligned parallel, i.e. in the same direction. This can be achieved for every pair simultaneously if *all* the spins point in the same direction:

$$\ldots \uparrow \uparrow \uparrow \uparrow \uparrow \ldots$$

which is clearly a *ferromagnetic* arrangement. In this case the energy is $E = NJS^2 = \frac{1}{4}NJ$ which is negative.

If $J > 0$ the lowest energy of a pair is when they are antiparallel. Provided N is even this can be achieved for all pairs on the chain by alternating:

$$\ldots \uparrow \; \downarrow \; \uparrow \downarrow \uparrow \downarrow \uparrow \downarrow \ldots$$
$$N{-}1 \; N \; 1 \; 2 \; 3 \; 4$$

which is an *anti*ferromagnetic arrangement. The energy is $E = -NJS^2 = -\frac{1}{4}NJ$.

Since each nearest-neighbour pair has its minimum energy configuration this arrangement minimises the overall energy and any other arrangement of the atoms would have higher energy.

Notice that rotation of *all* the spins by same amount does not affect the classical energy, i.e.

$$\ldots \nearrow \nearrow \nearrow \nearrow \nearrow \ldots \text{ or} \ldots \to \to \to \to \to \ldots$$

have the same energy as $\ldots \uparrow \uparrow \uparrow \uparrow \uparrow \ldots$ for the ferromagnetic arrangement. Similarly the antiferromagnetic arrangement can be rotated without changing the overall (classical) energy.

References

1. Cohen-Tannoudji, C., Diu, B., Laloë, F.: Quantum Mechanics I. Wiley Interscience, New York, NY (1977)
2. Kosterlitz, J.M., Thouless, D.J.: J. Phys. C Solid State Phys. **6**, 1181 (1973)

Chapter 3
Quantum Treatment of the Spin-$\frac{1}{2}$ Chain

Abstract For the spin-1/2 linear chain the simplest state is the fully aligned state, which is the ground state if the interaction is ferromagnetic. The ground state energy is easily calculated. The elementary excitations are states in which the spin of one atom is reversed, although the actual states are linear combinations of many of these states. Again the energies can be calculated exactly. For antiferromagnetic coupling these states, along with the aligned state, are still eigenstates albeit of much higher in energy than the actual ground state. Much information can also be obtained from a detailed study of the states with two reversals. Using these results it is possible to gradually increase the number of such reversals and still obtain exact eigenstates. This is the fundamental idea of the Bethe Ansatz, which is described in detail. When exactly half the atoms are reversed then the true antiferromagnetic ground state is obtained.

3.1 General Remarks

In the previous chapter we obtained some results using a classical treatment of the spin chain with the Hamiltonian Eq. (2.4). In this chapter we give a quantum mechanical treatment and obtain many more results, some of which have classical analogues which were not given there.

As a first step we construct a basis for the N spin-$\frac{1}{2}$ atoms. For a single atom a basis consists of the two states $\{|+\rangle, |-\rangle\}$. For the complete chain a basis would be all states of form

$$| + + - - + - - + - - - - + + \ldots\rangle$$
$$ 1 \ 2 \ 3 \ 4 \ 5$$

where the spin of each atom may be $+$ or $-$. Clearly the number of states in the basis is 2^N which is very large. We can now proceed to find the eigenstates of Eq. (2.4) using this basis.

First rewrite Eq. (2.4) as

$$\mathcal{H} = J \sum_{i=1}^{N} \left[S_i^z S_{i+1}^z + \frac{1}{2} \left(S_i^+ S_{i+1}^- + S_i^- S_{i+1}^+ \right) \right]. \tag{3.1}$$

Parkinson, J.B., Farnell, D.J.J.: *Quantum Treatment of the Spin-1/2 Chain*. Lect. Notes Phys. **816**, 21–38 (2010)
DOI 10.1007/978-3-642-13290-2_3

As before, we define the total z component as

$$S_T^z = \sum_j S_j^z.$$

Now clearly $[S_T^z, S_i^z S_{i+1}^z] = 0$ since only S^z operators are involved.

Also
$$
\begin{aligned}
[S_T^z, S_i^+ S_{i+1}^-] &= S_i^+[S_T^z, S_{i+1}^-] + [S_T^z, S_i^+]S_{i+1}^- \\
&= S_i^+[S_{i+1}^z, S_{i+1}^-] + [S_i^z, S_i^+]S_{i+1}^- \\
&= S_i^+(-S_{i+1}^-) + S_i^+ S_{i+1}^- = 0
\end{aligned}
$$

and similarly $[S_T^z, S_i^- S_{i+1}^+] = 0$. Hence S_T^z commutes with each term in $\mathcal{H}$ so

$$[S_T^z, \mathcal{H}] = 0.$$

This result is the same as in the two spin case and again we can choose the eigenstates of $\mathcal{H}$ to be eigenstates of S_T^z also.

Also, just as in the two spin case, because the Heisenberg exchange interaction is isotropic, it follows that if we define S_T^x and S_T^y similarly, each of these will also commute with $\mathcal{H}$. (This will not be true for other types of interaction.) Therefore the square of the total angular momentum (or spin)

$$\mathbf{S}_T^2 \equiv S_T^{x^2} + S_T^{y^2} + S_T^{z^2}$$

also commutes with $\mathcal{H}$. For example $[\mathbf{S}_T^2, \mathcal{H}] = 0$.

Hence for the Heisenberg Hamiltonian we can choose the eigenstates of $\mathcal{H}$ to be simultaneously eigenstates of $\mathbf{S}_T^2$ and any one of S_T^z, S_T^x or S_T^y. They cannot be eigenstates of all four operators since none of S_T^z, S_T^x or S_T^y commutes with each other. In practice we always choose them to be simultaneous eigenstates of $\mathbf{S}_T^2$ and S_T^z.

3.2 Aligned State

There is only one state in the basis with $S_T^z = N\frac{1}{2}$ namely the aligned state

$$|A\rangle \equiv |+++\cdots+++\rangle$$

which has all spins up. It follows that this must be an eigenstate of $\mathcal{H}$. ($\mathcal{H}$ commutes with S_T^z so there is no coupling to states with different S_T^z). In fact we can see this clearly by calculating

$$\mathcal{H}|A\rangle = J \sum_i \left[S_i^z S_{i+1}^z + \frac{1}{2} \left(S_i^+ S_{i+1}^- + S_i^- S_{i+1}^+ \right) \right] |A\rangle \qquad (3.2)$$

First note that $S_i^z|A\rangle = \frac{1}{2}|A\rangle$ since the ith atom is $|+\rangle$, and so $S_i^z S_{i+1}^z|A\rangle = \frac{1}{4}|A\rangle$.

Also $S_i^+|A\rangle = 0$ since the ith atom is in the state $|+\rangle$ and $S_{i+1}^+|A\rangle = 0$ since the $(i+1)$th atom is also in the state $|+\rangle$. Therefore, $S_i^+ S_{i+1}^-|A\rangle = 0$ and $S_i^- S_{i+1}^+|A\rangle = 0$. Thus,

$$\mathcal{H}|A\rangle = J \sum_i \left[\frac{1}{4}|A\rangle + 0|A\rangle \right] = \frac{JN}{4}|A\rangle.$$

This proves that the aligned state $|A\rangle$ is an eigenstate of $\mathcal{H}$ with eigenvalue $\frac{NJ}{4}$. It is useful to define $E_A \equiv \frac{NJ}{4}$, the energy eigenvalue of the aligned state.

If J is negative this is the actual ground state, and since all the atoms are parallel ('up') it is ferromagnetic. The eigenvalue E_A is the ground state energy of the system and it is exactly the same as in the classical case. It is important to notice that the ground state is highly degenerate. The same arguments show that the state with all atoms 'down' is also an eigenstate with eigenvalue E_A. In fact there is an eigenstate with this eigenvalue for *every* value of S_T^z, and this corresponds to the rotation without change of energy that we saw in the classical case.

If J is positive then $|A\rangle$ is actually the state of highest energy. It is still a valid eigenstate but obviously not the ground state. In fact as we shall see later the ground state is very complicated but is truly antiferromagnetic.

3.3 Single Deviation States

Now consider states with $S_T^z = \frac{N}{2} - 1$, i.e. one deviation from the aligned state. In the basis there are N states with a single spin in the $|-\rangle$ state. These are

$$
\begin{aligned}
|-++++\ldots++\rangle &\equiv |1\rangle \\
|+-+++\ldots++\rangle &\equiv |2\rangle \\
|++-++\ldots++\rangle &\equiv |3\rangle \quad \text{etc.}
\end{aligned}
$$

When we change a spin from $|+\rangle$ to $|-\rangle$ the value of S^z for that spin changes from $+\frac{1}{2}$ to $-\frac{1}{2}$, so the change is -1. Hence each of the above states has $S_T^z = \frac{N}{2} - 1$. There are no other states in the basis with this value of S_T^z.

Consider one of these states

$$|j\rangle \equiv |+++\ldots+-+\ldots++\rangle$$

$$\uparrow$$
$$j\text{th site}$$

and act upon it by terms in $\mathcal{H}$ given by Eq. (3.1).

$$
\begin{aligned}
S_i^z S_{i+1}^z |j\rangle &= \frac{1}{4}|j\rangle \quad \text{if } j \neq i,\ i+1 \\
&= -\frac{1}{4}|j\rangle \quad \text{if } j = i,\ i+1 .
\end{aligned}
$$

In $\sum_i S_i^z S_{i+1}^z$ there are two terms in which $i = j$ or $i+1 = j$, all the others $(N-2$ of these) satisfy the first condition. Hence

$$
\sum_i S_i^z S_{i+1}^z |j\rangle = \left[(N-2)\frac{1}{4} + 2\left(-\frac{1}{4}\right) \right] |j\rangle = \left(\frac{N}{4} - 1\right)|j\rangle .
$$

Now consider the effect of operator $S_i^+ S_{i+1}^-$ on $|j\rangle$.

$$
S_i^+ S_{i+1}^- |j\rangle = 0 \quad \text{unless } i = j \quad \text{since only the } j\text{th spin is 'down'}
$$

If $i = j$ then

$$
\begin{aligned}
S_j^+ S_{j+1}^- |j\rangle &= S_j^+ S_{j+1}^- |++\ldots+\underset{j}{-}+\ldots+\rangle \\
&= S_{j+1}^- |++\ldots++\underset{j}{+}+\ldots+\rangle \\
&= |++\ldots++\underset{j+1}{-}\ldots+\rangle = |j+1\rangle .
\end{aligned}
$$

Therefore $\sum_i S_i^+ S_{i+1}^- |j\rangle = |j+1\rangle$.
        Similarly

$$
\sum_i S_i^- S_{i+1}^+ |j\rangle = |j-1\rangle . \quad \text{(Coming from the term with } i+1 = j)
$$

Hence, using Eq. (3.1)

$$
\begin{aligned}
\mathcal{H}|j\rangle &= J\left[\left(\frac{N}{4} - 1\right)|j\rangle + \frac{1}{2}|j-1\rangle + \frac{1}{2}|j+1\rangle \right] \\
&= E_A |j\rangle + J\left[\frac{1}{2}|j-1\rangle + \frac{1}{2}|j+1\rangle - |j\rangle \right] . \quad (3.3)
\end{aligned}
$$

A true eigenstate state of $\mathcal{H}$ is constructed as a linear combination of the $|j\rangle$ by putting

$$\psi = \sum_j f_j |j\rangle \tag{3.4}$$

where the f_j are coefficients. The Schrödinger equation is

$$\mathcal{H}\psi = E\psi$$

and therefore

$$\sum_j f_j \mathcal{H}|j\rangle = E \sum_j f_j |j\rangle$$

$$\sum_j f_j E_A |j\rangle + \sum_j f_j J \left[\frac{1}{2}|j-1\rangle + \frac{1}{2}|j+1\rangle - |j\rangle \right] = E \sum_j f_j |j\rangle.$$

Now operate on the left by $\langle \ell |$. Since the basis is orthonormal

$$\langle \ell | j \rangle = \delta_{\ell j} \quad \text{etc.}$$

and therefore

$$J \left[\frac{1}{2}f_{\ell+1} + \frac{1}{2}f_{\ell-1} - f_\ell \right] = (E - E_A)f_\ell = \varepsilon f_\ell, \tag{3.5}$$

where $\varepsilon = E - E_A$ is the energy difference between this state and the aligned state.

This is a simple difference equation for the f_ℓ. The solutions have the form of plane waves

$$f_\ell = c_k e^{ik\ell} \quad (c_k \text{ constant}) \tag{3.6}$$

as can be seen by substituting in Eq. (3.5):

$$J \left[\frac{1}{2}c_k e^{ik(\ell+1)} + \frac{1}{2}c_k e^{ik(\ell-1)} - c_k e^{ik\ell} \right] = \varepsilon_k c_k e^{ik\ell}$$

where ε_k is the value of ε associated with k. Dividing by $c_k e^{ik\ell}$

$$J \left[\frac{1}{2}e^{ik} + \frac{1}{2}e^{-ik} - 1 \right] = \varepsilon_k$$

$$\text{and so} \quad \varepsilon_k = J(\cos k - 1). \tag{3.7}$$

Clearly we have solutions of (3.5) with the form (3.6) for any value of k. However, site $\ell + N$ is the same as site ℓ because of the periodic boundary conditions, and therefore

$$f_{\ell+N} = f_\ell.$$

Using (3.6)
$$c_k e^{ik(\ell+N)} = c_k e^{ik\ell}$$

so
$$e^{ikN} = 1 = e^{i2\pi\lambda} \quad \text{where } \lambda \text{ is an integer.}$$

It follows that k is given by

$$k = \lambda\frac{2\pi}{N} \quad \text{with} \quad 0 \le \lambda \le N-1.$$

There are N different eigenstates of this form corresponding to the N possible values of λ.

In summary we have found that eigenstates with $S_T^z = \frac{N}{2} - 1$ (i.e. with 1 net deviation from the aligned state) have the form

$$\psi_k = c_k \sum_\ell e^{ik\ell}|\ell\rangle \quad \text{where} \quad k = \lambda\frac{2\pi}{N}$$

with λ an integer, and the corresponding eigenvalue is $\varepsilon_k = J(\cos k - 1)$. c_k is an arbitrary constant. Normalising the eigenstates by putting $\langle \psi_k | \psi_k \rangle = 1$ gives

$$c_k = \frac{1}{\sqrt{N}}.$$

If J is negative then the ground state is the aligned ferromagnetic state and these states are the *elementary excitations* and are called 'spin-waves' or 'magnons'. The excitation energy ε_k is given by

$$\varepsilon_k = -J(1 - \cos k)$$

and since J is negative

$$\varepsilon_k = |J|(1 - \cos k) \quad \text{which is shown in Fig. 3.1}$$

The parameter k is often called a wave-vector even though in 1D it is a 'one-component vector', effectively a scalar.

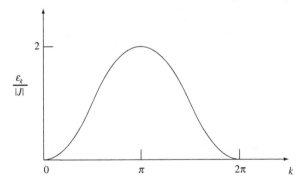

Fig. 3.1 Spectrum of elementary excitations (spin waves) of the 1D spin-$\frac{1}{2}$ chain with ferromagnetic isotropic nearest-neighbour Heisenberg exchange

3.4 Two Deviation States

Now consider states with $S_T^z = \frac{N}{2} - 2$, i.e. two deviations from the aligned state. As we shall see the eigenstates have the form of two spin waves which interact with each other. Sometimes the interaction is small and the energies of the spin waves are only slightly perturbed from those of free spin waves, together with small shifts in the k-vectors. In other cases the interaction is strong leading to a new type of state known as a bound state.

In the basis there are $\frac{N(N-1)}{2}$ states with two '$-$' spins. For example

$$|\underset{1 \ 2}{--} + + + + \cdots +\rangle \equiv |12\rangle$$

$$|\cdots + \underset{j_1}{-} + + + + \cdots + \underset{j_2}{-} + \ldots\rangle \equiv |j_1 j_2\rangle$$

Clearly $|j_1 j_2\rangle \equiv |j_2 j_1\rangle$ so to avoid overcounting we insist that $j_2 > j_1$. Note $j_2 \neq j_1$ since if we try to lower S^z for a spin-$\frac{1}{2}$ which is already in a '$-$' state we get 0 :

$$S_{j_1}^- |+\rangle = |-\rangle$$
$$(S_{j_1}^-)^2 |+\rangle = S_{j_1}^- |-\rangle = 0.$$

Now act on the basis states with the terms in $\mathcal{H}$.
Firstly consider $S_i^z S_{i+1}^z |j_1 j_2\rangle$:

a. $S_i^z S_{i+1}^z |j_1 j_2\rangle = \frac{1}{4}|j_1 j_2\rangle$ provided $j_1 \neq i, i+1$ and $j_2 \neq i, i+1$, i.e. neither of i nor $i+1$ is '$-$'.

b. $S_i^z S_{i+1}^z |j_1 j_2\rangle = -\frac{1}{4}|j_1 j_2\rangle$ if $i = j_1$ and $i+1 \neq j_2$ or
if $i \neq j_1$ and $i+1 = j_2$, i.e. one of $i, i+1$ is '$-$' but not both.

c. $S_i^z S_{i+1}^z |j_1 j_2\rangle = \frac{1}{4}|j_1 j_2\rangle$ if $i = j_1$ and $i+1 = j_2$, i.e. both i and $i+1$ are '$-$'.

There are thus two distinct cases after summing over i :

(a) j_1, j_2 adjacent i.e. $j_2 = j_1 + 1$ e.g. $|++\cdots+--+\ldots\rangle$ then

$$\sum_i S_i^z S_{i+1}^z |j_1 j_1 + 1\rangle$$
$$= \left[(N-3)\frac{1}{4} + 2\left(-\frac{1}{4}\right) + 1\left(\frac{1}{4}\right)\right]|j_1 j_1 + 1\rangle$$

where the first term comes from the case with $i, i+1$ both '+', the second from the case with one '+' and the other '−' and the third term from the case with both '−'. Hence

$$\sum_i S_i^z S_{i+1}^z |j_1 j_1 + 1\rangle = \left(\frac{N}{4} - 1\right) |j_1 j_1 + 1\rangle.$$

(b) j_2 not adjacent to j_1 $|++-++\cdots++-++\rangle$

$$\sum_i S_i^z S_{i+1}^z |j_1 j_2\rangle$$
$$= \left[(N-4)\frac{1}{4} + 4\left(-\frac{1}{4}\right)\right] |j_1 j_2\rangle = \left(\frac{N}{4} - 2\right) |j_1 j_2\rangle.$$

Now consider

$$S_i^+ S_{i+1}^- |j_1 j_2\rangle$$

This will be 0 unless $i = j_1$ or j_2. It will also be 0 if $i+1 = j_1$ or j_2.
    Again after summing over i there are two distinct cases:

(a) j_1, j_2 adjacent

$$\sum_i S_i^+ S_{i+1}^- |j_1 j_1 + 1\rangle = |j_1 j_1 + 2\rangle,$$

only one term in the sum giving a non-zero result.

(b) j_1, j_2 not adjacent

$$\sum_i S_i^+ S_{i+1}^- |j_1 j_2\rangle = |j_1 + 1, j_2\rangle + |j_1 j_2 + 1\rangle,$$

thus two terms giving non-zero results.
Similarly

$$\sum_i S_i^- S_{i+1}^+ |j_1 j_2\rangle = |j_1 - 1, j_1 + 1\rangle \quad \text{if } j_2 = j_1 + 1$$
$$= |j_1 - 1, j_2\rangle + |j_1, j_2 - 1\rangle \quad \text{otherwise.}$$

Hence, if $j_2 \neq j_1 + 1$ (and $j_1 \neq 1$, $j_2 \neq N$, due to periodic boundary conditions) then

$$\mathcal{H}|j_1 j_2\rangle = J\left(\frac{N}{4} - 2\right)|j_1 j_2\rangle$$
$$+\frac{J}{2}\Big(|j_1 - 1, j_2\rangle + |j_1 + 1, j_2\rangle + |j_1, j_2 - 1\rangle + |j_1, j_2 + 1\rangle\Big),$$

while, if $j_2 = j_1 + 1$ (or $j_1 = 1$ and $j_2 = N$, again due to periodic boundary conditions) then

$$\mathcal{H}|j_1 j_1 + 1\rangle = J\left(\frac{N}{4} - 1\right)|j_1 j_1 + 1\rangle + \frac{J}{2}\Big(|j_1 - 1, j_1 + 1\rangle + |j_1, j_1 + 2\rangle\Big).$$

The actual eigenstate is a linear combination of the form

$$\psi = \sum_{j_1 j_2} f_{j_1 j_2}|j_1 j_2\rangle \qquad j_2 > j_1,$$

so the Schrödinger equation $\mathcal{H}\psi = E\psi$ gives

$$\sum_{j_1=1}^{N-1} \sum_{j_2=j_1+1}^{N} f_{j_1 j_2}\mathcal{H}|j_1 j_2\rangle = \sum_{j_1}\sum_{j_2} f_{j_1 j_2} E|j_1 j_2\rangle.$$

Multiplying on the left by $\langle \ell_1 \ell_2|$ and using

$$\langle \ell_1 \ell_2|j_1 j_2\rangle = \delta_{\ell_1 j_1}\delta_{\ell_2 j_2}$$
$$\langle \ell_1 \ell_2|j_1 - 1 j_2\rangle = \delta_{\ell_1 j_1-1}\delta_{\ell_2 j_2} \qquad \text{etc.}$$

gives eventually two equations

$$(E_A - E - 2J)f_{\ell_1 \ell_2} + \frac{J}{2}\Big(f_{\ell_1-1\ell_2} + f_{\ell_1+1\ell_2} + f_{\ell_1\ell_2-1} + f_{\ell_1\ell_2+1}\Big) = 0 \quad (3.8)$$

provided $\ell_2 \neq \ell_1 + 1$, and

$$(E_A - E - J)f_{\ell_1 \ell_1+1} + \frac{J}{2}\Big(f_{\ell_1-1\ell_1+1} + f_{\ell_1\ell_1+2}\Big) = 0. \qquad (3.9)$$

Equation (3.9) acts as a sort of 'boundary condition' to (3.8). It differs from (3.8) in that there are only two 'incorrect' bonds (i.e. of the form $+-$ or $-+$ rather than $++$) instead of four giving J instead of $2J$ in the first term. Furthermore, terms where the lowering operator acts twice at the same site are missing, as these cannot occur for spin$-\frac{1}{2}$.

Again these are difference equations and we can easily find a solution of (3.8) of the form

$$f_{\ell_1,\ell_2} = e^{ik_1\ell_1}e^{ik_2\ell_2} \tag{3.10}$$

Proof After substituting (3.10) into (3.8) and then dividing by $e^{ik_1\ell_1}e^{ik_2\ell_2}$ we get

$$(-\varepsilon - 2J) + \frac{J}{2}\left(e^{-ik_1} + e^{ik_1} + e^{-ik_2} + e^{ik_2}\right) = 0$$

where $\varepsilon \equiv E - E_A$. Therefore

$$\varepsilon = J(\cos k_1 + \cos k_2 - 2) \tag{3.11}$$

Provided this is satisfied then f_{ℓ_1,ℓ_2} as given is a solution. Note that this solution has an energy ε which consists of sum of two energies of same form as for a single deviation state $\varepsilon_k = J(\cos k - 1)$. However, as we shall see, the values of k_1, k_2 are not the same as the values which occur in ε_k.

We must still satisfy the 'boundary condition' Eq. (3.9). For a given value of ε given by (3.11), then clearly both

$$f_{\ell_1,\ell_2} = e^{ik_1\ell_1}e^{ik_2\ell_2}$$

and

$$f'_{\ell_1,\ell_2} = e^{ik_2\ell_1}e^{ik_1\ell_2}$$

satisfy (3.8). The most general solution of (3.8) is thus

$$f_{\ell_1,\ell_2} = c_1 e^{ik_1\ell_1}e^{ik_2\ell_2} + c_2 e^{ik_2\ell_1}e^{ik_1\ell_2}. \tag{3.12}$$

Only the ratio of c_1 and c_2 is important as the absolute values can be determined by normalisation. Put $\frac{c_1}{c_2} = e^{i\phi}$ (we allow ϕ to be complex so there is no loss of generality here) and then write

$$f_{\ell_1,\ell_2} = C\left(e^{ik_1\ell_1}e^{ik_2\ell_2}e^{i\phi/2} + e^{ik_1\ell_2}e^{ik_2\ell_1}e^{-i\phi/2}\right) \tag{3.13}$$

where C is an overall normalisation constant.

Substituting this into (3.9), then, after dividing by $Ce^{ik_1\ell_1}e^{ik_2\ell_1}e^{-i\phi/2}$, we get

$$(-\varepsilon - J)\left(e^{ik_2}e^{i\phi} + e^{ik_1}\right) + \frac{J}{2}\left(e^{-ik_1}e^{ik_2}e^{i\phi} + e^{-ik_2}e^{ik_1} + e^{i2k_2}e^{i\phi} + e^{i2k_1}\right) = 0.$$

It is convenient to introduce $x_1 = e^{ik_1}$, $x_2 = e^{ik_2}$, and $y = e^{i\phi}$ so that

$$(-\varepsilon - J)(x_2 y + x_1) + \frac{J}{2}\left(x_2^2 y + x_1^2 + x_1^{-1} x_2 y + x_1 x_2^{-1}\right) = 0. \qquad (3.14)$$

Now use the earlier result that

$$\varepsilon = J(\cos k_1 + \cos k_2 - 2)$$

$$\text{so} \quad -\varepsilon - J = J(1 - \cos k_1 - \cos k_2)$$

$$= J\left(1 - \frac{x_1}{2} - \frac{x_1^{-1}}{2} - \frac{x_2}{2} - \frac{x_2^{-1}}{2}\right).$$

Substituting into (3.14) and cancelling J gives

$$Py + Q = 0 \qquad (3.15)$$

where

$$P = \left(1 - \frac{x_1}{2} - \frac{x_1^{-1}}{2} - \frac{x_2}{2} - \frac{x_2^{-1}}{2}\right) x_2 + \frac{1}{2} x_2^2 + \frac{1}{2} x_1^{-1} x_2$$

$$= x_2 - \frac{x_1 x_2}{2} - \frac{1}{2}.$$

Q is the same except that $x_1 \leftrightarrow x_2$,

$$\text{so} \quad Q = x_1 - \frac{x_1 x_2}{2} - \frac{1}{2}.$$

From (3.15) $\quad y = -\dfrac{Q}{P}$

$$\text{so} \quad \frac{y+1}{y-1} = \frac{\left(-\frac{Q}{P} + 1\right)}{\left(-\frac{Q}{P} - 1\right)} = \frac{Q - P}{Q + P}.$$

$$\text{Now} \quad Q - P = x_1 - x_2$$

$$\text{and} \quad Q + P = x_1 + x_2 - x_1 x_2 - 1$$

which can be rewritten as :

$$Q - P = x_1 - x_2 = \frac{1}{2}\Big[(1 - x_2)(1 + x_1) - (1 - x_1)(1 + x_2)\Big]$$

and

$$Q + P = x_1 + x_2 - x_1 x_2 - 1 = -(1 - x_1)(1 - x_2).$$

Therefore

$$\frac{y+1}{y-1} = \frac{1}{2}\left[\frac{(1-x_2)(1+x_1) - (1-x_1)(1+x_2)}{-(1-x_1)(1-x_2)}\right]$$

$$2\left(\frac{y+1}{y-1}\right) = \left(\frac{x_1+1}{x_1-1}\right) - \left(\frac{x_2+1}{x_2-1}\right).$$

Now note that $i\left(\dfrac{x_1+1}{x_1-1}\right) = i\left(\dfrac{e^{ik_1}+1}{e^{ik_1}-1}\right) = \cot\dfrac{k_1}{2}$.

Similarly $i\left(\dfrac{x_2+1}{x_2-1}\right) = \cot\dfrac{k_2}{2}$ and $i\left(\dfrac{y+1}{y-1}\right) = \cot\dfrac{\phi}{2}$,

so that finally

$$2\cot\frac{\phi}{2} = \cot\frac{k_1}{2} - \cot\frac{k_2}{2}. \tag{3.16}$$

This equation is an important relation between the k's and ϕ. However it does not tell us the allowed values of k_1 and k_2. For this we need other equations, which come from the periodic boundary conditions.

Clearly $f_{\ell_1+N,\ell_2+N} = f_{\ell_1,\ell_2}$ and from (3.10) this implies

$$e^{ik_1 N} e^{ik_2 N} = 1$$

so $(k_1 + k_2)N = \lambda 2\pi$ where λ is an integer,

and therefore $k_1 + k_2 = \lambda\dfrac{2\pi}{N}.$ \tag{3.17}

$k_1 + k_1$ is the *total* wavevector, and the fact that this has to be an integer multiple of $2\pi/N$ merely reflects the translational symmetry.

More interestingly, however, we also have

$$f_{\ell_2,\ell_1+N} = f_{\ell_1,\ell_2}$$

since the $(\ell_1 + N)$'th site is the same as the ℓ_1'th site and our convention is that the second subscript is always greater than the first. Now (3.12) gives

$$e^{ik_1\ell_2} e^{ik_2(\ell_1+N)} e^{i\phi/2} + e^{ik_2\ell_2} e^{ik_1(\ell_1+N)} e^{-i\phi/2} = e^{ik_1\ell_1} e^{ik_2\ell_2} e^{i\phi/2} + e^{ik_2\ell_1} e^{ik_1\ell_2} e^{-i\phi/2}.$$

For this to be true for all ℓ_1, ℓ_2, the coefficients of $e^{ik_1\ell_1} e^{ik_2\ell_2}$ and also of $e^{ik_2\ell_1} e^{ik_1\ell_2}$ must be the same on both sides. The first of these gives

$$e^{ik_1 N} e^{-i\phi/2} = e^{i\phi/2}$$

$$\therefore \quad e^{i(k_1 N - \phi)} = 1$$

$$\therefore \quad k_1 N - \phi = \lambda_1 2\pi \quad \text{where } \lambda_1 \text{ is an integer.}$$

$$\text{and so} \quad k_1 = \lambda_1 \frac{2\pi}{N} + \frac{\phi}{N}. \tag{3.18}$$

Similarly the second gives

$$k_2 = \lambda_2 \frac{2\pi}{N} - \frac{\phi}{N} \quad \text{where } \lambda_2 \text{ is an integer.} \tag{3.19}$$

Note that the sum of (3.18) and (3.19) is (3.17), so that only two of these are strictly needed.

Conclusion The eigenstates with two deviations from the fully aligned state have the form:

$$\psi = \sum_{\ell_1 \ell_2} f_{\ell_1, \ell_2} |\ell_1 \ell_2\rangle \quad (\ell_2 > \ell_1) \tag{3.20}$$

$$f_{\ell_1, \ell_2} = C\left(e^{ik_1\ell_1} e^{ik_2\ell_2} e^{i\phi/2} + e^{ik_2\ell_1} e^{ik_1\ell_2} e^{-i\phi/2}\right) \tag{3.21}$$

where k_1, k_2 and ϕ satisfy the three equations

$$2\cot\phi/2 = \cot\frac{k_1}{2} - \cot\frac{k_2}{2} \tag{3.22}$$

$$k_1 = \lambda_1 \frac{2\pi}{N} + \frac{\phi}{N} \tag{3.23}$$

$$k_2 = \lambda_2 \frac{2\pi}{N} - \frac{\phi}{N}, \tag{3.24}$$

and the energy of this eigenstate is

$$\varepsilon = J(\cos k_1 + \cos k_2 - 2).$$

It is important to note here that k_1 and k_2 are *not* integer multiples of $\frac{2\pi}{N}$ and so the energy is *not* the same as for two single deviation states. In fact k_1 and k_2 may not even be real, since ϕ does not have to be real.

3.4.1 Form of the States

There are two types of two-deviation states as follows.

a. *Class C* These occur when $|\lambda_1 - \lambda_2| \geq 2$. It can be shown that in this case the ϕ are real, and therefore k_1, k_2 are also real.
   Since the k_1, k_2 are real the magnitude of each of the two terms in f_{ℓ_1, ℓ_2} is always 1, independent of the values of ℓ_1, ℓ_2. We say that the state consists of two 'free' deviations, or two 'free' spin waves or two 'free' magnons. Note they are 'free' but they still interact with each other and this causes the shift in k_1, k_2 from the values allowed for isolated spin waves. In other words they are not independent.

For a given value of total wavevector $k = k_1 + k_2$, the energy can be written as

$$\varepsilon = -J[2 - \cos k_1 - \cos(k - k_1)].$$

Choosing the ferromagnetic case with $J = -|J|$, this can be written as

$$\varepsilon = |J| \left[2 - 2\cos\frac{k}{2} \cos\left(k_1 - \frac{k}{2}\right) \right].$$

k_1 can take values in the range $0 \leq k_1 \leq 2\pi$ so the second cosine can effectively have any value between -1 and $+1$. This means that these states with two 'free' magnons have energies for a given k bounded by $|J|(2 - 2\cos\frac{k}{2})$ below and $|J|(2 + 2\cos\frac{k}{2})$ above. This is indicated in Fig. 3.2 by the curves marked P and Q. As $N \to \infty$ these states form a continuum.

b. *Class A/B* These occur for $\lambda_1 = \lambda_2$ and $|\lambda_1 - \lambda_2| = 1$, although there is no significant difference between these two cases. It can be shown that in these cases the ϕ and therefore k_1, k_2 are complex in such a way that $k_2 = k_1^*$.

Now the magnitude of the terms in f_{ℓ_1,ℓ_2} does depend on ℓ_1 and ℓ_2. In fact one of the terms in (3.12) diverges as $|\ell_2 - \ell_1| \to \infty$. This is unphysical and the coefficient of this term must be chosen to be zero. The other term then tends to 0 as $|\ell_2 - \ell_1| \to \infty$, which means that the two deviations are more likely to be found close to each other than further apart. For this reason such a state is called a bound state of the two deviations or of two spin waves. The two deviations form a complex which can then travel freely along the chain. Another term used to describe this state is a '2-string'.

Again it can be shown that energy of the bound state lies above that of two 'free' states with the same total k if $J > 0$, i.e. the antiferromagnetic case, but below them if $J < 0$, the ferromagnetic case.

The spectrum of the bound states can be obtained by writing

$$f_{\ell_1,\ell_2} = c_1 e^{i\frac{k}{2}(\ell_1+\ell_2)} e^{-g(\ell_2-\ell_1)}. \tag{3.25}$$

where k is the total wave-vector and g is a positive real constant. Comparing with the earlier form, Eq. (3.12), the second term has been omitted by choosing $c_2 = 0$ since if ℓ_1 and ℓ_2 are interchanged the term with g would diverge as $\ell_2 - \ell_1 \to \infty$, as mentioned earlier. The earlier form, Eq. (3.12), is valid for any value of N whereas this form, Eq. (3.25), is an approximation valid for large N as it requires $\ell_2 - \ell_1 << N$. Because of the exponential decay this is valid for the bound states.

Substituting (3.25) into Eqs. (3.8) and (3.9) gives

$$(E_A - E - 2J) + \frac{J}{2} \left(e^{-i\frac{k}{2}}e^{-g} + e^{i\frac{k}{2}}e^{g} + e^{-i\frac{k}{2}}e^{g} + e^{i\frac{k}{2}}e^{-g} \right) = 0 \tag{3.26}$$

and

$$(E_A - E - J) + \frac{J}{2}\left(e^{i\frac{k}{2}}e^{-g} + e^{-i\frac{k}{2}}e^{-g}\right) = 0 \qquad (3.27)$$

which together yield

$$e^{-g} = \frac{1}{2}\left(e^{i\frac{k}{2}} + e^{-i\frac{k}{2}}\right) = \cos\frac{k}{2}. \qquad (3.28)$$

Finally substituting back into (3.27) gives

$$E - E_A = -\frac{J}{2}(1 - \cos k)$$

This curve has the same form as the free magnon but with a coefficient which is half that of the free magnon. It is indicated by curve R in Fig. 3.2.

We see from (3.12) and (3.25) that for $c_2 = 0$, $e^{i\frac{k}{2}(l_1+l_2)}e^{-g(l_2-l_1)} \equiv e^{ik_1l_1}e^{ik_2l_2}$, so that $k_1 = k/2 - ig$ and $k_2 = k/2 + ig$. Hence this special solution is also the same as (3.23) and (3.24) provided

$$k_1 = \lambda_1\frac{2\pi}{N} + \frac{\phi}{N} = \frac{k}{2} - ig,$$

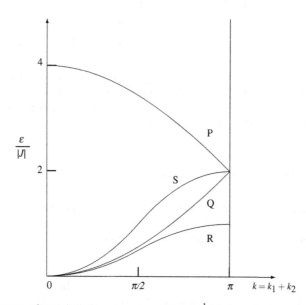

Fig. 3.2 Spectrum of two-deviation states in a 1D spin-$\frac{1}{2}$ chain with ferromagnetic isotropic nearest-neighbour Heisenberg exchange. P, Q are the top and bottom of the continuum of two 'free'-magnon states. R is the single branch of two-magnon bound states and S is the single spin-wave spectrum

$$\text{and} \quad k_2 = \lambda_2 \frac{2\pi}{N} - \frac{\phi}{N} = \frac{k}{2} + ig,$$

which requires $\quad \phi = (\lambda_2 - \lambda_1)\pi - igN.$

3.5 Three Deviation States

Now consider states with $S_T^z = \frac{N}{2} - 3$, i.e. three deviations from the aligned state. If we have three deviations then similar equations can be written down for the coefficients $f_{\ell_1\ell_2\ell_3}$ as were written for the two-deviation coefficients $f_{\ell_1\ell_2}$. Again there is one basic equation for the case where none of ℓ_1, ℓ_2, ℓ_3 are neighbours. This has a solution of the form

$$f_{\ell_1\ell_2\ell_3} = e^{ik_1\ell_1} e^{ik_2\ell_2} e^{ik_3\ell_3}$$

with $\varepsilon = -J(3 - \cos k_1 - \cos k_2 - \cos k_3)$. Clearly this is the same energy as three free magnons with wave vectors k_1, k_2, k_3, although, just as for the case of two deviations, these wave vectors will not be the same as for free magnons. For the same value of ε any permutation of k_1, k_2, k_3 will be a solution, so we expect in general

$$\begin{aligned} f_{\ell_1\ell_2\ell_3} = {}& A_1 e^{i(k_1\ell_1 + k_2\ell_2 + k_3\ell_3)} + A_2 e^{i(k_2\ell_1 + k_3\ell_2 + k_1\ell_3)} \\ &+ A_3 e^{i(k_3\ell_1 + k_1\ell_2 + k_2\ell_3)} + B_1 e^{i(k_1\ell_1 + k_3\ell_2 + k_2\ell_3)} \\ &+ B_2 e^{i(k_3\ell_1 + k_2\ell_2 + k_1\ell_3)} + B_3 e^{i(k_2\ell_1 + k_1\ell_2 + k_3\ell_3)} \end{aligned} \quad (3.29)$$

The ratio of the coefficients are determined by 'boundary condition' equations when two or three of ℓ_1, ℓ_2, ℓ_3 are neighbours. These equations are somewhat lengthy and are not given here. Again it is useful to introduce phase factors and write $A_p = Ce^{i\Phi_p/2}$ and $B_p = Ce^{-i\Phi_p/2}$ for $p = 1, 2, 3$.

3.5.1 Bethe Ansatz for $S_T^Z = \frac{N}{2} - 3$

Bethe [1] suggested (The German word 'ansatz' has no exact equivalent in English but roughly means 'starting point'), and subsequently proved, that the phase factors Φ_p should have the form of a sum of '2-body' phase factors of the sort we saw earlier. Specifically that

$$\left. \begin{aligned} \Phi_1 &= \phi_{12} + \phi_{13} + \phi_{23} \\ \Phi_2 &= \phi_{23} + \phi_{21} + \phi_{31} \\ \Phi_3 &= \phi_{31} + \phi_{32} + \phi_{12} \end{aligned} \right\} \begin{array}{l} \text{Note that the ordering of the} \\ \text{subscripts here is the same as the ordering} \\ \text{of the } k\text{'s in the corresponding term.} \end{array}$$

where the ϕ_{ij} satisfy the same equations as before, namely

$$2 \cot \frac{\phi_{ij}}{2} = \cot \frac{k_i}{2} - \cot \frac{k_j}{2} . \tag{3.30}$$

These equations ensure that $\phi_{ji} = -\phi_{ij}$ and hence the phases of B_p are the negative of those of A_p.

Periodic boundary conditions applied to $f_{\ell_1 \ell_2 \ell_3}$ require

$$\left. \begin{array}{l} Nk_1 = 2\pi \lambda_1 + \phi_{12} + \phi_{13} \\ Nk_2 = 2\pi \lambda_2 + \phi_{21} + \phi_{23} \\ Nk_3 = 2\pi \lambda_3 + \phi_{31} + \phi_{32} \end{array} \right\}$$

where $\lambda_1, \lambda_2, \lambda_3$ are integers.

Again we find that if $|\lambda_i - \lambda_j| \geq 2$ for all i, j then the solutions have three real values of k_i. These are states with three 'free' spin waves. If some $|\lambda_i - \lambda_j| = 0$ or 1 then we get bound states or mixtures of bound and 'free'. A state may consist either of 1 'free' and 1 bound pair of deviations in which case one k is real and the other two complex or of a single bound state of 3 deviations in which case all the k are complex. The form of the bound state of 3 deviations is a generalisation of Eq. (3.25) but will not be given here.

3.6 States with an Arbitrary Number of Deviations

In general we may have r deviations from the fully aligned state, i.e. $S_T^Z = \frac{N}{2} - r$. Bethe showed that for r deviations the solution would be a generalised version of the three-deviation form Eq. (3.29). In particular there are now r values of k_i, denoted $k_1, k_2, \ldots, k_r$.

There will be $M = r!$ permutations of these which we label $P_1 \ldots P_M$. Let the ith k in the pth permutation be k_i^p, then

$$f_{\ell_1 \ldots \ell_r} = \sum_{p=1}^{M} A_p \, e^{i(k_1^p \ell_1 + k_2^p \ell_2 \ldots k_r^p \ell_r)}$$

Note that the summation here is over the M permutations. The Bethe ansatz now requires

$$A_p = C \exp \left[\sum_{i=1}^{r} \sum_{j=i+1}^{r} \phi_{ij}^p \right]$$

where $\phi_{ij}^p = \pm \phi_{ij}$, the negative sign occurring if the order of k_i, k_j in the pth permutation is reversed. For example, referring back to the three-deviation case, $\Phi_2 = \phi_{23} - \phi_{12} - \phi_{13}$. In addition Eq. (3.30) still applies.

This form of $f_{\ell_1 \dots \ell_r}$ is clearly rather complicated, but in practice we do not need to use it (for some things!). Normally, the actual equations we work with are the following:

$$\text{the energy equation } \varepsilon \ = \ J \sum_{i=1}^{r} (\cos k_i - 1) \tag{3.31}$$

and the two equations relating k and ϕ

$$N k_i = 2\pi \lambda_i + \sum_{j \neq i} \phi_{ij} \tag{3.32}$$

$$2 \cot \frac{\phi_{ij}}{2} = \cot \frac{k_i}{2} - \cot \frac{k_j}{2}. \tag{3.33}$$

These are the equations that form the Bethe Ansatz and were first written down by Bethe in 1931 [1]. Bethe showed that they are correct for the $S = \frac{1}{2}$ Heisenberg model. They do not apply for $S > \frac{1}{2}$.

The fact that the Bethe Ansatz works for $S = \frac{1}{2}$ means that it is an example of what is called an *integrable* system. The $S > \frac{1}{2}$ systems are not integrable. Unfortunately there is no general way of predicting which systems are integrable. Each case has to be demonstrated directly.

Clearly integrable systems are special cases and in general most systems are not integrable. Nevertheless since many exact results are available for integrable systems they are extremely important and they have greatly increased our understanding of many-body systems.

Reference

1. Bethe, H.A.: Z. Phys. **71**, 205–226 (1931)

Chapter 4
The Antiferromagnetic Ground State

Abstract This chapter gives the mathematical details of the calculation of the ground state energy of the spin-1/2 linear chain with antiferromagnetic nearest neighbour exchange. Although the form of the ground-state wave function had been given by Bethe using the Bethe Ansatz, as described in the previous chapter, it was several years before Hulthén was able to use it to calculate the ground-state energy. The procedure involves setting up an integral equation for a function f. Although f does not have a simple physical significance, the complete wave function is made up of a superposition of phase-shifted plane waves with wave vector k. f is related to the rate of change of the density of the distribution k. Once the fundamental integral equation has been derived it is solved by Fourier transform. Finally the solution f is used to find the antiferromagnetic ground-state energy.

4.1 The Fundamental Integral Equation

Although Bethe had given the wave functions in 1931 they were in a rather difficult formal form, involving sums over permutations. The first really useful result based on Bethe's work was by Hulthén in 1938 [1].

As we saw for two spins, if J is −ve, then the ground state is ferromagnetic and completely aligned. This state is also very simple from a quantum mechanical point of view as the eigenstate is one of the basis states. However, for J +ve the ground state is *anti*ferromagnetic and even for two spins is more complicated.

When we considered states with two deviation from the aligned state in the N atom chain we found two types of states

a. 2 'free' deviations
b. bound pair.

If J is positive the bound states lie above the free ones in energy, so the lowest state is the class C type for which k_1, k_2 and ϕ_{12} are all real. These are states for which $|\lambda_1 - \lambda_2| \geq 2$. For r deviations the same result applies. The lowest states are those for which all the k_i and ϕ_{ij} are real. Again these are class C type and have $|\lambda_i - \lambda_j| \geq 2$ for all i, j.

Parkinson, J.B., Farnell, D.J.J.: *The Antiferromagnetic Ground State*. Lect. Notes Phys. **816**, 39–47 (2010)
DOI 10.1007/978-3-642-13290-2_4

The *classical* antiferromagnetic ground state is (choosing the direction of alignment of the spins to be parallel and antiparallel to the z-axis)

$$\ldots + \ - \ + \ - \ + \ - \ + \ - \ + \ - \ldots$$

which is a state with $\frac{N}{2}$ atoms reversed. The corresponding quantum mechanical state, known as the Néel state,

$$| + \ - \ + \ - \quad \ldots \ \rangle$$

is a basis state but is not an eigenstate. Nevertheless the true ground state will not be orthogonal to this state and since it is a state with $S_T^Z = 0$, i.e. $r = \frac{N}{2}$, then the ground state will have $S_T^z = 0$ and $r = \frac{N}{2}$ also.

The equations we need to solve are

$$k_i = \lambda_i \frac{2\pi}{N} + \frac{1}{N} \sum_{j \neq i} \phi_{ij} \quad (\lambda_i \text{ integer}) \tag{4.1}$$

and

$$2 \cot \frac{\phi_{ij}}{2} = \cot \frac{k_i}{2} - \cot \frac{k_j}{2} \tag{4.2}$$

Clearly $\lambda_i + N$ is equivalent to λ_i (since adding N to λ_i merely increases k_i by 2π). Hence we can restrict λ_i to be an integer in the range $0 \leq \lambda_i \leq N - 1$ without loss of generality. The antiferromagnetic ground state has $r = \frac{N}{2}$, i.e. $\frac{N}{2}$ values of λ_i, and is class C so we require $|\lambda_i - \lambda_j| \geq 2$ for all i, j. There are two possible choices:

$$\{\lambda_i\} = \{0, 2, 4, \ldots, N - 2\}$$
$$\text{OR} \quad \{\lambda_i\} = \{1, 3, 5, \ldots, N - 1\}$$

The first of these has $\lambda_1 = 0$ and it can be shown that this gives a state in which $S_T = 1$ with $S_T^Z = 0$, while the second choice gives a state with $S_T = 0$ and $S_T^Z = 0$. The second of these is the correct one since the first is degenerate with other states with $S_T = 1$ and $S_T^Z = \pm 1$ which lie higher in energy.

Now because the λ_i are uniformly spaced, we can introduce a new variable

$$x_i = \frac{2i - 1}{N} = \frac{\lambda_i}{N},$$

which becomes a continuous variable in the limit $N \to \infty$, running from 0 to 1. Likewise we can regard the k_i as forming a continuous set, i.e.

$$x_i \equiv \frac{\lambda_i}{N} \xrightarrow[N \to \infty]{} x \qquad k_i \xrightarrow[N \to \infty]{} k(x) \qquad 0 \leq x \leq 1.$$

In Eq. (4.2) we have k_j as well as k_i. k_j is associated with λ_j and we introduce

$$y_j \equiv \frac{\lambda_j}{N} \xrightarrow[N \to \infty]{} y \qquad\qquad k_j \xrightarrow[N \to \infty]{} k(y) \qquad\qquad 0 \le y \le 1,$$

so that Eq. (4.2) becomes

$$2 \cot \frac{1}{2} \phi(x, y) = \cot \frac{k(x)}{2} - \cot \frac{k(y)}{2} \tag{4.3}$$

and we shall restrict ϕ to lie in the range $-\pi \le \phi \le \pi$. Note that the y here is not the same as the y in the previous chapter.

Equation (4.1) becomes

$$k(x) = 2\pi x + \frac{1}{2} \int_0^1 \phi(x, y)dy \tag{4.4}$$

(the factor $\frac{1}{2}$ comes because the λ_j are separated by 2). Note that the range of $k(x)$ is $0 \le k(x) \le 2\pi$, unlike that of ϕ. It can be shown that if $x < y$ then $k(x) < k(y)$ as one would expect from Eq. (4.4) provided the integral is well-behaved.

These are the integral equations solved by Huthén. The treatment given here is based on that of Mattis [2] with important extra details. First make the substitutions

$$\cot \frac{k(x)}{2} = \xi(x)$$

$$\cot \frac{k(y)}{2} = \eta(y),$$

so that $2 \cot \frac{1}{2} \phi(x, y) = \xi(x) - \eta(y)$.

When $x = y$, $\cot \frac{1}{2}\phi = 0$, so $\phi = \pm\pi$ and at this point the value of ϕ jumps from $-\pi$ to π. For this reason we divide the integral in (4.4) into two sections:

$$k(x) = 2\pi x + \frac{1}{2} \int_0^x \phi(x, y)dy + \frac{1}{2} \int_x^1 \phi(x, y)dy \tag{4.5}$$

In the first integral $y < x$ therefore $k(y) < k(x)$. This implies, as shown in Fig. 4.1, and recalling that $0 \le k(x), k(y) \le 2\pi$, that $\cot \frac{k(y)}{2} > \cot \frac{k(x)}{2}$ and therefore

$$\cot \frac{\phi}{2} = \frac{1}{2} \left[\cot \frac{k(x)}{2} - \cot \frac{k(y)}{2} \right] < 0.$$

As shown in Fig. 4.1, if $\cot \frac{\phi}{2} < 0$ and $-\pi \le \phi \le \pi$ we must have $\phi < 0$. Similarly in the second integral we have $\phi > 0$.

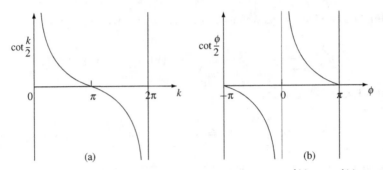

Fig. 4.1 (a) Plot of $\cot\frac{k}{2}$ versus k showing that if $k(y) < k(x)$ then $\cot\frac{k(y)}{2} > \cot\frac{k(x)}{2}$. (b) Plot of $\cot\frac{\phi}{2}$ versus ϕ showing that $\cot\frac{\phi}{2}$ has the same sign as ϕ

Now differentiate (4.5) with respect to x

$$\frac{dk}{dx} = 2\pi + \frac{1}{2}\phi_1(x, x) + \frac{1}{2}\int_0^x \frac{\partial\phi}{\partial x}dy$$
$$- \frac{1}{2}\phi_2(x, x) + \frac{1}{2}\int_x^1 \frac{\partial\phi}{\partial x}dy$$

where $\phi_1(x, x) = \lim_{y\to x-} \phi(x, y)$ in the first integral in Eq. (4.5)

$$= -\pi$$

and $\phi_2(x, x) = \lim_{y\to x+} \phi(x, y)$ in the second integral in Eq. (4.5)

$$= +\pi$$

so

$$\frac{dk}{dx} = 2\pi + \frac{1}{2}(-\pi) - \frac{1}{2}(\pi) + \frac{1}{2}\int_0^x \frac{\partial\phi}{\partial x}dy + \frac{1}{2}\int_x^1 \frac{\partial\phi}{\partial x}dy$$
$$= \pi + \frac{1}{2}\int_0^1 \frac{\partial\phi}{\partial x}dy.$$

Next we introduce the functions

$$f(\xi) = -\frac{dx}{d\xi}; \quad f(\eta) = -\frac{dy}{d\eta}$$

then

$$\frac{\partial\phi}{\partial x} = \frac{\partial\phi}{\partial\xi}\frac{d\xi}{dx} = -\frac{1}{f(\xi)}\frac{\partial\phi}{\partial\xi}.$$

But $\phi = 2\cot^{-1}\left[\frac{1}{2}(\xi - \eta)\right],$

therefore $\dfrac{\partial \phi}{\partial \xi} = 2 \dfrac{-1}{1 + \frac{1}{4}(\xi - \eta)^2}\dfrac{1}{2}.$

Also $dy = \dfrac{dy}{d\eta} d\eta = -f(\eta)\, d\eta,$

therefore $\dfrac{dk}{dx} = \pi + \dfrac{1}{2}\displaystyle\int \left(-\dfrac{1}{f(\xi)}\right)\dfrac{(-1)}{1 + \frac{1}{4}(\xi - \eta)^2}(-f(\eta)\, d\eta)$

$$= \pi - \frac{1}{2f(\xi)}\int \frac{f(\eta)d\eta}{1 + \frac{1}{4}(\xi - \eta)^2}.$$

This is now an integral over η instead of y. As y goes from 0 to 1, $k(y)$ goes from 0 to 2π (not proved here but can be shown from (4.4)) so $\cot\frac{k(y)}{2}$ goes from $+\infty$ to $-\infty$. Therefore

$$\frac{dk}{dx} = \pi + \frac{1}{2f(\xi)}\int_{-\infty}^{\infty} \frac{f(\eta)\, d\eta}{1 + \frac{1}{4}(\xi - \eta)^2}.$$

Finally $\dfrac{dk}{dx} = \dfrac{dk}{d\xi}\dfrac{d\xi}{dx} = -\dfrac{1}{f(\xi)}\dfrac{dk}{d\xi}$

and since $\xi = \cot\dfrac{k(x)}{2}$, $k(x) = 2\cot^{-1}\xi$ and $\dfrac{dk}{d\xi} = \dfrac{-2}{1 + \xi^2}.$

Therefore

$$\frac{2}{1 + \xi^2}\frac{1}{f(\xi)} = \pi + \frac{1}{2f(\xi)}\int_{-\infty}^{\infty}\frac{f(\eta)d\eta}{1 + \frac{1}{4}(\xi - \eta)^2}$$

i.e. $f(\xi) = \dfrac{2}{\pi(1 + \xi^2)} - \dfrac{1}{2\pi}\displaystyle\int_{-\infty}^{\infty}\dfrac{f(\eta)d\eta}{1 + \frac{1}{4}(\xi - \eta)^2}.$ (4.6)

This is the fundamental integral equation for $f(\xi)$ (see, Mattis [2] 5.168).

4.2 Solution of the Fundamental Integral Equation

Linear integral equations in which the kernel is a function of $(\xi - \eta)$ only, can be solved by the Fourier Transform. We define

$$F(q) = \int_{-\infty}^{\infty} f(\theta)e^{iq\theta}d\theta$$

with inverse

$$f(\theta) = \frac{1}{2\pi} \int_{-\infty}^{\infty} F(q)e^{-iq\theta} dq.$$

Substituting in Eq. (4.6) gives

$$\frac{1}{2\pi} \int_{-\infty}^{\infty} F(q)e^{-iq\xi} dq = \frac{2}{\pi(1+\xi^2)}$$
$$- \frac{1}{2\pi} \int_{-\infty}^{\infty} \left(\frac{1}{2\pi} \int_{-\infty}^{\infty} F(q)e^{-iq\eta} dq \right) \frac{d\eta}{1 + \frac{1}{4}(\xi - \eta)^2}.$$

Now

$$\int_{-\infty}^{\infty} \frac{e^{-iq\eta} d\eta}{1 + \frac{1}{4}(\xi - \eta)^2} = e^{-iq\xi} \int_{-\infty}^{\infty} \frac{e^{iq(\xi-\eta)} d\eta}{1 + \frac{1}{4}(\xi - \eta)^2}.$$

so substituting $z = \frac{1}{2}(\xi - \eta)$, with $d\eta = -2dz$, gives

$$= 2e^{-iq\xi} \int_{-\infty}^{\infty} \frac{e^{2iqz} dz}{1 + z^2}$$

$$= 2e^{-iq\xi} \pi e^{-2q}.$$

Therefore

$$\frac{1}{2\pi} \int_{-\infty}^{\infty} F(q)e^{-iq\xi} dq = \frac{2}{\pi(1+\xi^2)} - \frac{1}{2\pi} \int_{-\infty}^{\infty} F(q)e^{iq\xi} e^{-2q} dq.$$

Multiply both sides by $e^{iq'\xi}$ and integrate $\int_{-\infty}^{\infty} d\xi$

$$F(q') = \frac{2}{\pi} \int_{-\infty}^{\infty} \frac{e^{iq'\xi} d\xi}{(1+\xi^2)} - e^{-2q'} F(q')$$

$$(1 + e^{-2q'})F(q') = \frac{2}{\pi} \pi e^{-q'}$$

$$\text{therefore } F(q') = \frac{1}{\cosh q'}$$

$$\text{i.e.} \quad F(q) = \text{sech}(q).$$

Finally

$$f(\xi) = \frac{1}{2\pi} \int_{-\infty}^{\infty} \text{sech}(q) e^{-iq\xi} dq \; = \; \frac{1}{2} \text{sech}\left(\frac{\pi\xi}{2}\right).$$

4.3 The Ground State Energy

We can now calculate the ground state energy of the infinite chain. For a finite chain
with N atoms we have seen that the ground state energy is given by

$$\varepsilon = J \sum_i (\cos k_i - 1).$$

In the limit $N \to \infty$ we again make the changes

$$k_i \to k(x) \quad \text{and} \quad \sum_{i=1}^{N/2} \longrightarrow \frac{N}{2} \int_0^1 dx$$

so that

$$\varepsilon \; = \; J \frac{N}{2} \int_0^1 [\cos k(x) - 1] \, dx.$$

Also $\cos k(x) - 1 = -2\sin^2 \dfrac{k}{2}$

$$= -\frac{2}{\text{cosec}^2 \frac{k}{2}} \; = \; \frac{-2}{1 + \cot^2 \frac{k}{2}} \; = \; \frac{-2}{1 + \xi^2}$$

while $dx = \dfrac{dx}{d\xi} d\xi \; = \; -f(\xi) d\xi.$

Therefore

$$\varepsilon = -\frac{JN}{2} \int_{-\infty}^{\infty} \frac{2f(\xi)d\xi}{1 + \xi^2}$$

$$= -JN \int_{-\infty}^{\infty} \left(\frac{1}{2\pi} \int_{-\infty}^{\infty} F(q) e^{-iq\xi} dq \right) \frac{d\xi}{(1 + \xi^2)}$$

Doing the ξ integral

$$= -JN\frac{1}{2\pi}\int_{-\infty}^{\infty} F(q)\pi e^{-|q|}dq$$

$$= -\frac{JN}{2}\int_{-\infty}^{\infty} \frac{2}{(e^q+e^{-q})}e^{-|q|}dq$$

$$= -JN2\int_{0}^{\infty} \frac{e^{-q}}{e^q+e^{-q}}dq$$

$$= -JN\ln 2 \qquad \text{(using the substitution } z = e^{-q}.)$$

Therefore

$$E = \varepsilon + E_A = -JN\ln 2 + \frac{JN}{4} \qquad (4.7)$$

$$= -0.443147J \quad \text{to six decimal places.} \qquad (4.8)$$

This is a very famous result due to Hulthén and Bethe. It is one of the outstanding achievements of modern theoretical physics.

As well as this exact result for the ground state energy of the spin-$\frac{1}{2}$ chain with isotropic Heisenberg exchange, the Bethe Ansatz [3] has been used to obtain many other results for spin-$\frac{1}{2}$ chains. Some of these are as follows:

a. It may be extended to the case of anisotropic exchange. Firstly to the XXZ-model in which $S^x S^x$ and $S^y S^y$ terms are equal but the $S^z S^z$ terms are different, by Orbach [4] and Walker [5] and later to the more general XYZ-model in which all three terms are different, by Baxter [6].
b. For the XXZ-model the correlation functions behave as

$$\left|\left\langle S_i^z S_{i+R}^z\right\rangle\right| = \text{const. as } R \to \infty$$

for the anisotropic model ($\Delta > 1$) and

$$\left|\left\langle S_i^z S_{i+R}^z\right\rangle\right| \propto \frac{1}{R} \text{ as } R \to \infty$$

for the isotropic model ($\Delta = 1$) showing that there is no long range order for $\Delta = 1$. These and other important results for the correlation function of this model were obtained by a completely different method, the inverse scattering method, by Bogoliubov et al. [7]
c. It was extended to include the effects of a magnetic field by Griffiths [8]
d. Results for the elementary excitations (antiferromagnetic magnons) can be found as described in the next chapter.
e. The thermodynamics, i.e. the properties at non-zero temperature, have been studied by Yang and Yang [9–11], Takahashi and Suzuki [12], Gaudin [13] and Klümper [14]

References

1. Hulthen, L.: Ark. Mater. Astron. Fys. A **26**, 11–116 (1938)
2. Mattis, D.C.: The Theory of Magnetism I, Springer, Berlin (1981, 1988)
3. Bethe, H.A.: Z. Phys. **71**, 205–226 (1931)
4. Orbach, R.: Phys. Rev. **112**, 309–316 (1958)
5. Walker, L.R.: Phys. Rev. **116**, 1089–1090 (1959)
6. Baxter, R.J.: Ann. Phys. (New York) **70**, 323–337 (1972)
7. Bogoliubov, N.M., Izergin, A.G., Korepin, V.E.: Nucl. Phys. **B275**, 687–705 (1986)
8. Griffiths, R.B.: Phys. Rev. **133**, A768–A775 (1964)
9. Yang, C.N., Yang, C.P.: Phys. Rev. **150**, 321–327 (1966)
10. Yang, C.N., Yang, C.P.: Phys. Rev. **150**, 327–339 (1966)
11. Yang, C.N., Yang, C.P.: Phys. Rev. **151**, 258–264 (1966)
12. Takahashi, M., Suzuki, M.: Prog. Theor. Phys. **48**, 2187–2209 (1972)
13. Gaudin, M.: Phys. Rev. Lett. **26**, 1301–1304 (1971)
14. Klümper, A.: Europhys. Lett. **9**, 815–820 (1989)

Chapter 5
Antiferromagnetic Spin Waves

Abstract The ground state of the antiferromagnetic spin-1/2 linear chain is a made up of a linear combination of basis states with exactly half ($N/2$) the spins reversed. The elementary excitations from the ground state have $N/2 \pm 1$ spins reversed. To calculate the energies of these states, des Cloiseaux and Pearson used a modified version of the same method that Hulthén had used for the ground state energy. The modifications are quite significant and involve some rather different mathematical techniques. These are described in this chapter. The result is not a single branch of excitations as a function of the total wave vector, but rather a continuum of states. The lower boundary of this continuum has a simple sine-wave form, which is similar to that obtained using conventional antiferromagnetic spin-wave theory (covered in a later chapter). We finish this chapter with a discussion of the somewhat unexpected behaviour of the excitation spectrum when an external magnetic field is present.

5.1 The Basic Formalism

Bethe [1] gave the wave functions for the 1D $S = \frac{1}{2}$ Heisenberg chain in 1931 and the ground state energy was evaluated by Hulthén in 1938 [2]. In 1962 Des Cloiseaux and Pearson [3] obtained the exited states or elementary excitations, the antiferromagnetic spin waves. In this chapter we follow their treatment closely with some extra details. Note that the treatment given here determines the energies of the elementary excitations exactly. In Chap. 8 an approximate theory of the same excitations is given which is applicable much more generally than the $S = \frac{1}{2}$ Heisenberg chain.

The lowest-lying excited states (the elementary excitations) have $S_T = 1$ and $S_T^z = 0, \pm 1$. Once again, within the $S_T^z = 1$ subspace the lowest lying states are class C: states which are all or partly bound, with complex values of k_i, lie higher in energy.

For a state with $S_T^z = 1$ we require $\dfrac{N}{2} - 1$ deviations from the fully aligned state, and hence $\dfrac{N}{2} - 1$ values of k_i, with corresponding values of λ_i. For class C we require that the λ_i are separated by at 2 or more and are chosen from the integers

Parkinson, J.B., Farnell, D.J.J.: *Antiferromagnetic Spin Waves*. Lect. Notes Phys. **816**, 49–59 (2010)
DOI 10.1007/978-3-642-13290-2_5   © Springer-Verlag Berlin Heidelberg 2010

$1, 2, 3, \ldots, N - 2, N - 1$. λ_i must not be chosen to be zero since this gives the $S_T^z = 1$ component of a multiplet with $S_T > 1$.

There are clearly many ways to do this. One way is to choose

$$\lambda = 1, 3, 5, \ldots, \Lambda, \Lambda + 4, \Lambda + 6, \ldots, N - 3, N - 1 \qquad (5.1)$$

where Λ is an odd integer. Here we have a single gap of size 4. Alternatively we may have two gaps of size 3

$$\lambda = 1, 3, 5, \ldots, \Lambda_1, \Lambda_1 + 3, \Lambda_1 + 5, \ldots, \Lambda_2, \Lambda_2 + 3, \Lambda_2 + 5, \ldots, N - 3, N - 1$$
$$(5.2)$$

where Λ_1 is an odd integer and $\Lambda_2 \geq \Lambda_1 + 3$ is an even integer. One could also have 'gaps' at the beginning

$$\lambda = 3, 5, 7, \ldots, N - 5, N - 3, N - 1 \qquad (5.3)$$

or

$$\lambda = 2, 4, 6, \ldots, \Lambda_2, \Lambda_2 + 3, \Lambda_2 + 5, \ldots, N - 3, N - 1 \qquad (5.4)$$

or similar 'gaps' at the end.

$$\lambda = 1, 3, 5, 7, \ldots, N - 5, N - 3 \qquad (5.5)$$

or

$$\lambda = 1, 3, 5, \ldots, \Lambda_1, \Lambda_1 + 3, \Lambda_1 + 5, \ldots, N - 4, N - 2 \qquad (5.6)$$

These latter four can be regarded as special cases of Eq. (5.1) and Eq. (5.2), but in fact they are important as we shall see.

All these possibilities lead to a large number of states, forming a continuum in the limit $N \to \infty$. The lowest states for a given k were determined by Des Cloiseaux and Pearson [3] by studying numerically finite size chains with $N \leq 16$. They found that for $-\pi < k < 0$ or equivalently $\pi < k < 2\pi$ the choice that gives the lowest energy is given by Eq. (5.4), while for $0 < k < \pi$ the correct choice is Eq. (5.6). We shall follow des Cloiseaux and Pearson and present the calculation for $-\pi < k < 0$.

Two other preliminaries are necessary. Firstly it is convenient to work in the $S_T^z = 0$ subspace. As noted before an $S_T = 1$ state with $S_T^z = 0$ has an extra $\lambda = 0$ added since there need to be $\frac{N}{2}$ deviations, and this corresponds to an extra $k_i = 0$. Hence, writing $\Lambda_2 = 2n$, the set of λ is

$$\lambda = 0, 2, 4, 6, \ldots, 2n, 2n + 3, \ldots, N - 3, N - 1 \qquad (5.7)$$

Secondly the total wave-vector

$$k = \sum_{i=1}^{N/2} k_i = \sum_{i=1}^{N/2} \left(\frac{2\pi}{N} \lambda_i + \sum_{j=1}^{N/2}{}' \phi_{ij} \right) = \frac{2\pi}{N} \sum_{i=1}^{N/2} \lambda_i$$

since for every ϕ_{ij} in this sum there is an equal and opposite ϕ_{ji}. The prime on the second summation indicates $j \neq i$.

Using $1 + 3 + 5 + \cdots + (N - 3) + (N - 1) = \frac{N^2}{4}$ and noting that the set of λ in Eq. (5.7) differs from this by the first n terms being reduced by 1 we have

$$k = \frac{2\pi}{N} \left(\frac{N^2}{4} - n \right) = \frac{2\pi}{N}(-n) \quad \mathrm{mod}\ 2\pi,$$

which is strictly only true if N is a multiple of 4 but this is not a significant restriction in the limit $N \to \infty$. Thus

$$n = N|k|/2\pi. \tag{5.8}$$

As in the previous chapter, we now pass to the continuum limit by writing $x_i = \frac{2i-1}{N}$, which becomes a *continuous* variable in the limit $N \to \infty$, running from 0 to 1. Note that $x_i \neq \frac{\lambda_i}{N}$ now as the λ are not evenly spaced. In fact the λ satisfy

$$\lambda_i = \frac{2i - 2}{N} = \frac{x_i}{N} - \frac{1}{N} \quad \text{for } i \leq n, \tag{5.9}$$

$$\lambda_i = \frac{2i - 1}{N} = \frac{x_i}{N} \quad \text{for } i > n. \tag{5.10}$$

In the large N limit we define

$$x_i \equiv \frac{2i - 1}{N} \xrightarrow[N \to \infty]{} x; \qquad \frac{\lambda_i}{N} \xrightarrow[N \to \infty]{} \lambda(x); \qquad k_i \xrightarrow[N \to \infty]{} k(x); \qquad 0 \leq x \leq 1$$

The $i \leq n$ in Eq. (5.9) becomes $x < \frac{|k|}{\pi}$ and $i > n$ in Eq. (5.10) becomes $x > \frac{|k|}{\pi}$ and these two equations can be written together as

$$\lambda(x) = x - \left(\frac{1}{N} \right) \Theta \left(\frac{|k|}{\pi} - x \right) \tag{5.11}$$

where Θ is the step function.

The calculation given here differs from that in the previous chapter only in the choice of the λ_i. The other equations from the Bethe method are unchanged, namely

$$2 \cot \frac{1}{2} \phi(x, y) = \cot \frac{k(x)}{2} - \cot \frac{k(y)}{2}, \tag{5.12}$$

$$k(x) = 2\pi \lambda(x) + \frac{1}{2} \int_0^1 \phi(x, y)\, dy, \tag{5.13}$$

and

$$\varepsilon = -\frac{JN}{2} \int_0^1 [1 - \cos k(x)]\, dx, \tag{5.14}$$

and again we take $-\pi \le \phi \le \pi$.

Substituting for $\lambda(x)$ from Eq. (5.11) gives

$$k(x) = 2\pi \left[x - \left(\frac{1}{N} \right) \Theta \left(\frac{|k|}{\pi} - x \right) \right] + \frac{1}{2} \int_0^1 \phi(x, y)\, dy, \tag{5.15}$$

and splitting the integral into two parts for the same reason as before

$$k(x) = 2\pi \left[x - \left(\frac{1}{N} \right) \Theta \left(\frac{|k|}{\pi} - x \right) \right]$$

$$+ \frac{1}{2} \int_0^x \phi(x, y)\, dy + \frac{1}{2} \int_x^1 \phi(x, y)\, dy. \tag{5.16}$$

Now differentiate with respect to x and note that $\frac{d}{dx} \Theta \left(\frac{|k|}{\pi} - x \right) = -\delta \left(\frac{|k|}{\pi} - x \right)$.

$$\frac{dk}{dx} = 2\pi \left[1 + \frac{1}{N} \delta \left(\frac{|k|}{\pi} - x \right) \right] + \frac{1}{2} \phi_1(x, x) + \frac{1}{2} \int_0^x \frac{\partial \phi}{\partial x} dy$$

$$- \frac{1}{2} \phi_2(x, x) + \frac{1}{2} \int_x^1 \frac{\partial \phi}{\partial x} dy$$

where

$$\phi_1(x, x) = \lim_{y \to x-} \phi(x, y) = -\pi,$$

and

$$\phi_2(x, x) = \lim_{y \to x+} \phi(x, y) = +\pi \quad \text{as before.}$$

Thus, putting $x_0 = \dfrac{|k|}{\pi}$,

$$\frac{dk}{dx} = 2\pi \left[1 + \frac{1}{N} \delta(x_0 - x) \right] + \frac{1}{2}(-\pi) - \frac{1}{2}(\pi) + \frac{1}{2} \int_0^x \frac{\partial \phi}{\partial x} dy + \frac{1}{2} \int_x^1 \frac{\partial \phi}{\partial x} dy$$

$$= \pi + \frac{2\pi}{N} \delta(x_0 - x) + \frac{1}{2} \int_0^1 \frac{\partial \phi}{\partial x} dy.$$

Using the substitutions

$$\cot \frac{k(x)}{2} = \xi(x)$$

$$\cot \frac{k(y)}{2} = \eta(y),$$

so that $2 \cot \frac{1}{2} \phi(x, y) = \xi(x) - \eta(y)$, and defining

$$f(\xi) = -\frac{dx}{d\xi}; \quad f(\eta) = -\frac{dy}{d\eta}$$

then

$$\frac{\partial \phi}{\partial x} = \frac{\partial \phi}{\partial \xi} \frac{d\xi}{dx} = -\frac{1}{f(\xi)} \frac{\partial \phi}{\partial \xi}.$$

But

$$\phi = \cot^{-1} \left[\frac{1}{2}(\xi - \eta) \right],$$

therefore

$$\frac{\partial \phi}{\partial \xi} = \frac{-1}{1 + \frac{1}{4}(\xi - \eta)^2} \frac{1}{2}.$$

Also

$$dy = \frac{dy}{d\eta} d\eta = -f(\eta) d\eta,$$

$$\therefore \quad \frac{dk}{dx} = \pi + \frac{2\pi}{N} \delta(x_0 - x) + \frac{1}{2} \int \left(-\frac{1}{f(\xi)} \right) \frac{(-1)(-f(\eta) d\eta)}{1 + \frac{1}{4}(\xi - \eta)^2}$$

$$= \pi + \frac{2\pi}{N} \delta(x_0 - x) - \frac{1}{2f(\xi)} \int \frac{f(\eta) d\eta}{1 + \frac{1}{4}(\xi - \eta)^2}.$$

This is now an integral over η instead of y. As y goes from 0 to 1, $k(y)$ goes from 0 to 2π (not strictly proved here but can be shown from (5.15)) so $\cot \frac{k(y)}{2}$ goes from $+\infty$ to $-\infty$.

Putting $\xi_0 = \cot \frac{k(x_0)}{2}$ and using the standard relation for change of variable

$$\delta(x_0 - x) = \delta(\xi_0 - \xi) \frac{d\xi}{dx} = \delta(\xi_0 - \xi) \frac{-1}{f(\xi)}$$

and noting that $\delta(\xi_0 - \xi)\frac{-1}{f(\xi)} = \delta(\xi - \xi_0)\frac{1}{f(\xi)}$ since $f(\xi) = -f(-\xi)$, gives

$$\frac{dk}{dx} = \pi + \frac{2\pi}{Nf(\xi)}\delta(\xi - \xi_0) - \frac{1}{2f(\xi)}\int_{-\infty}^{\infty}\frac{f(\eta)\,d\eta}{1 + \frac{1}{4}(\xi - \eta)^2}.$$

Finally $\quad \dfrac{dk}{dx} = \dfrac{dk\,d\xi}{d\xi\,dx} = -\dfrac{1}{f(\xi)}\dfrac{dk}{d\xi}$

and since $\xi = \cot\dfrac{k(x)}{2}$, $k(x) = 2\cot^{-1}\xi$ and $\dfrac{dk}{d\xi} = \dfrac{-2}{1 + \xi^2}$.

Therefore

$$\frac{2}{1+\xi^2}\frac{1}{f(\xi)} = \pi + \frac{2\pi}{Nf(\xi)}\delta(\xi - \xi_0) - \frac{1}{2f(\xi)}\int_{-\infty}^{\infty}\frac{f(\eta)d\eta}{1 + \frac{1}{4}(\xi - \eta)^2} \qquad (5.17)$$

i.e. $\quad f(\xi) = \dfrac{2}{\pi(1+\xi^2)} + \dfrac{1}{2\pi}\int_{-\infty}^{\infty}\dfrac{f(\eta)d\eta}{1 + \frac{1}{4}(\xi - \eta)^2} - \dfrac{2}{N}\delta(\xi - \xi_0).$ $\qquad (5.18)$

Notice that the only difference between this equation and the corresponding $f(\xi)$ in the ground state calculation of the previous chapter is the extra delta function at the end. This comes from the choice of λ in Eqs. (5.9) and (5.10).

The method of solution of this equation is the same as in the previous chapter, using the Fourier Transform. We define

$$F(q) = \int_{-\infty}^{\infty} f(\theta)e^{iq\theta}\,d\theta$$

with inverse

$$f(\theta) = \frac{1}{2\pi}\int_{-\infty}^{\infty} F(q)e^{-iq\theta}\,dq.$$

Substituting in Eq. (5.18) gives

$$\frac{1}{2\pi}\int_{-\infty}^{\infty} F(q)e^{-iq\xi}\,dq = \frac{2}{\pi(1+\xi^2)} - \frac{2}{N}\delta(\xi - \xi_0)$$

$$+ \frac{1}{2\pi}\int_{-\infty}^{\infty}\frac{d\eta}{1 + \frac{1}{4}(\xi - \eta)^2}\frac{1}{2\pi}\int_{-\infty}^{\infty} F(q)e^{-iq\eta}dq.$$

Now

$$\int_{-\infty}^{\infty}\frac{e^{-iq\eta}d\eta}{1 + \frac{1}{4}(\xi - \eta)^2} = e^{-iq\xi}\int_{-\infty}^{\infty}\frac{e^{iq(\xi-\eta)}d\eta}{1 + \frac{1}{4}(\xi - \eta)^2}.$$

so substituting $z = \dfrac{1}{2}(\xi - \eta)$, with $d\eta = -2dz$, gives

$$= -2e^{-iq\xi} \int_{-\infty}^{\infty} \frac{e^{2iqz}dz}{1+z^2}$$

$$= -2e^{-iq\xi} \pi e^{-2q}.$$

Therefore

$$\frac{1}{2\pi} \int_{-\infty}^{\infty} F(q)e^{-iq\xi}dq = \frac{2}{\pi(1+\xi^2)} - \frac{2}{N}\delta(\xi - \xi_0) - \frac{1}{2\pi} \int_{-\infty}^{\infty} F(q)e^{iq\xi}e^{-2q}dq.$$

Multiply both sides by $e^{iq'\xi}$ and integrate $\int_{-\infty}^{\infty} d\xi$

$$F(q') = \frac{2}{\pi} \int_{-\infty}^{\infty} \frac{e^{iq'\xi}d\xi}{(1+\xi^2)} - e^{-2q'}F(q') - \frac{2}{N}e^{iq'\xi_0}$$

$$(1 + e^{-2q'})F(q') = \frac{2}{\pi}\pi e^{-q'} - \frac{2}{N}e^{iq'\xi_0}$$

therefore

$$F(q') = \frac{1}{\cosh q'} - \frac{2e^{iq'\xi_0}}{N(1+e^{-2q'})}$$

As before the energy relative to the aligned state is

$$\varepsilon = -\frac{JN}{2} \int_{-\infty}^{\infty} \frac{2f(\xi)d\xi}{1+\xi^2} \tag{5.19}$$

$$= -JN\frac{1}{2\pi} \int_{-\infty}^{\infty} F(q)\pi e^{-|q|}dq$$

$$= -\frac{JN}{2} \int_{-\infty}^{\infty} \frac{2}{(e^q + e^{-q})}e^{-|q|}dq + J \int_{-\infty}^{\infty} \frac{e^{-|q|}e^{iq\xi_0}}{1+e^{-2q}}dq$$

$$= -JN2 \int_{0}^{\infty} \frac{e^{-q}}{e^q + e^{-q}}dq + \frac{J}{2} \int_{-\infty}^{\infty} e^{iq\xi_0} \operatorname{sech} q \, dq$$

$$= -JN \ln 2 + \frac{J\pi}{2} \operatorname{sech}(\pi \xi_0/2) \tag{5.20}$$

Clearly the first term is the ground state energy and the second is the additional energy of the elementary excitations above the ground state.

Finally we need to relate ξ_0 to the wave-vector k. If we ignore terms of order $\frac{1}{N}$ in Eq. (5.15) we obtain the corresponding equation from the previous chapter. The relation between x and $k(x)$ in this calculation differs only to order $\frac{1}{N}$ from that in the previous so we can use the result obtained there:

$$-\frac{dx}{d\xi} = f(\xi) = \frac{1}{2}\,\text{sech}\left(\frac{\pi\xi}{2}\right)$$

When $x = 0$, $k(x) = 0$ so $\xi = \cot\frac{k(x)}{2} = +\infty$.
Hence

$$\int_0^{x_0} dx = -\int_\infty^{\xi_0} \frac{1}{2}\,\text{sech}\left(\frac{\pi\xi}{2}\right) d\xi = \frac{1}{\pi}\left[\cot^{-1}\sinh\left(\frac{\pi\xi}{2}\right)\right]_\infty^{\xi_0}$$

$$\therefore \quad x_0 = \frac{1}{\pi}\cot^{-1}\left[\sinh\left(\frac{\pi\xi_0}{2}\right)\right].$$

To order $\dfrac{1}{N}$, $\pi x_0 = k$ so

$$\cot(k) = \sinh\left(\frac{\pi\xi_0}{2}\right)$$

and

$$\text{sech}\left(\frac{\pi\xi_0}{2}\right) = \frac{1}{\sqrt{1 + \sinh^2(\frac{\pi\xi_0}{2})}} = \frac{1}{\sqrt{1 + \cot^2(k)}}$$

$$= \frac{1}{\sqrt{\text{cosec}^2(k)}} = \sin k$$

Substituting into Eq. (5.20) gives

$$\varepsilon = -JN\,\ln 2 + \frac{J\pi}{2}\sin k \tag{5.21}$$

This is the energy relative to the aligned state so the excitation energy, the energy above the ground state is

$$E_k = \frac{J\pi}{2}\sin k \tag{5.22}$$

This is the exact result obtained by Des Cloiseaux and Pearson [3] for the energy of the antiferromagnetic spin-waves or magnons in the spin-$\frac{1}{2}$ chain with isotropic Heisenberg exchange.

It should be noted that the particular choice of λ given by Eqs. (5.9) and (5.10) is not the only possible one for class C states with $N/2 - 1$ deviations from the aligned state, i.e. with $S_T^z = 1$. Des Cloiseaux and Pearson showed numerically that it gives the lowest state for any particular k. The other choices lead to a continuum of states, bounded above by $E_k^{\text{max}} = J\pi\sin\frac{k}{2}$ [4]. The energies of the class C states are shown in Fig. 5.1. One should also remember that there are numerous states which

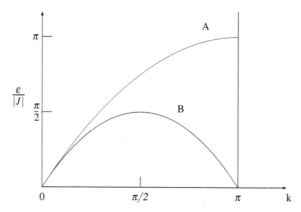

Fig. 5.1 Elementary excitation energies in a 1D spin-$\frac{1}{2}$ chain with antiferromagnetic isotropic nearest-neighbour Heisenberg exchange. There is a continuum of states from the lower boundary B to the upper boundary A. The states on the lower boundary B are the antiferromagnetic spin waves

are not pure class C but have one or more bound multiplets and which lie in general at higher energy than the class C states.

Nevertheless the des Cloiseaux and Pearson result is extremely important and in fact the response to a probe, for example by neutron scattering, is strongest at this lower boundary [5, 6].

5.2 Magnetic Field Behaviour

Finally in this chapter we discuss the elementary excitations in the presence of a magnetic field H applied in the z-direction. In this case the Hamiltonian for the chain becomes

$$\mathcal{H} = J \sum_{i=1}^{N} \mathbf{S}_i \cdot \mathbf{S}_{i+1} - g\beta H \sum_{i=1}^{N} S_i^z \tag{5.23}$$

where g is the Landé g-factor mentioned in Chap. 1 and β is the Bohr magneton, the magnetic dipole moment associated with one unit of angular momentum.

For antiferromagnetic coupling, $J > 0$, the classical ground state consists of two sublattices, exactly as in the zero field case, but now the orientation of the spins on one sublattice is at an angle θ to the z-axis and on the other sublattice at an angle $-\theta$. In fact the ground state is the same as for an interacting pair of spins for which

$$\mathcal{H} = J\mathbf{S}_1 \cdot \mathbf{S}_2 - \frac{1}{2} g\beta H (S_1^z + S_2^z)$$
$$= JS^2 \cos 2\theta - g\beta HS \cos\theta \tag{5.24}$$

and minimising with respect to θ gives

$$JS^2 2 \sin 2\theta - g\beta H S \sin\theta = 0$$

from which $\quad \cos\theta = \dfrac{H}{K} \quad$ where $K = \dfrac{4JS}{g\beta}$

and so $\quad S^z = S\cos\theta = \dfrac{H}{2K}.$

Clearly the largest value of S^z is $\frac{1}{2}$ and this is reached when $H = K$. That this is the classical ground state for the whole system follows from the fact that this choice gives the minimum energy for every pair separately.

It should be pointed out that the magnetic fields needed in practice to produce values of θ significantly less than 90°, the zero-field value, are much too high to be realised experimentally, so for the present this remains a theoretical exercise. However, it is of interest because it is a striking example of the difference between classical and quantum behaviour.

The existence of two sublattices implies that the underlying periodicity of the system is $2a$, rather than a, the lattice spacing, which we shall take to be unity. The Brillouin zone (BZ) should then have a periodicity in k-space of $\frac{2\pi}{2a} = \pi$. This is precisely the periodicity observed in the elementary excitations described in the previous section, in the absence of a magnetic field.

The quantum treatment of the elementary excitations was first carried out by Ishimura and Shiba [7], see also [8], using the same Bethe Ansatz method as is used for the elementary excitations in zero field, but with a modified choice of the λ. The result is that the periodicity in k-space of the BZ is quite different from the classical periodicity. In fact the periodicity changes smoothly from π at $H = 0$ where the magnetisation per site $\langle S^z \rangle = 0$, to 0 at $H = K$ where $\langle S^z \rangle = \frac{1}{2}$. The spectrum is shown for various values of the magnetisation in Fig. 5.2. Note that the periodicity for the fully aligned state is actually 2π rather than zero and the behaviour when

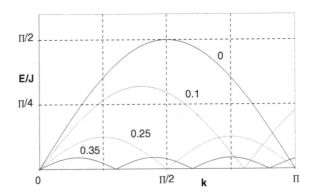

Fig. 5.2 The spectrum of elementary excitations for an antiferromagnetic spin 1/2 chain in a magnetic field. The curves are labelled by the magnetisation per site $\langle S^z \rangle$, which is zero in zero field and $\frac{1}{2}$ for fields $H \geq K$ where all the spins are aligned parallel to the field

the number of reversed atoms is small, of order $1/N$, i.e. when $\langle S^z \rangle \to \frac{1}{2}$, has to be handled differently.

This clearly indicates that the underlying magnetic structure does not have a periodicity of $2a$ in real space like the classical structure. A simple model [9] which fits the data quite well is one in which there are no sublattices as such but rather the spins all align parallel or antiparallel to the z-axis. Then as the magnetic field increases the antiparallel spins become fewer in number but regularly spaced, resulting in a steadily increasing periodicity in real space. This periodicity varies with the magnetisation in a way that precisely matches the observed variation of the periodicity of the BZ in k-space.

References

1. Bethe, H.A.: Z. Phys. **71**, 205–226 (1931)
2. Hulthen, L.: Ark. Mater. Astron. Fys. A **26**, 11–116 (1938)
3. Des Cloiseaux, J., Pearson, J.J.: Phys. Rev. **128**, 2131 (1962)
4. Yamada, T.: Prog. Theor. Phys. **41**, 880 (1969)
5. Karbach, M., et al.: Phys. Rev. B **55**, 12510 (1997).
6. Karbach, M., et al.: Phys. Rev. B **55**, 2131 (1962).
7. Ishimura, N., Shiba, H.: Prog. Theor. Phys. **57**, 1862–1873 (1977)
8. Aghahosseini, H., Parkinson, J.B.: J. Phys. C Solid State Phys. **13**, 651–665 (1980)
9. Aghahosseini, H., Parkinson, J.B.: J. Phys. C Solid State Phys. **14**, 425–437 (1981)

Chapter 6
The XY Model

Abstract The XY Model is another linear chain of spin-1/2 atoms but with a different type of exchange interaction in which only the x-components and y-components of the spins are involved, and with unequal weights. This system has interesting and unusual features and it is exactly soluble i.e. integrable. However the mathematical techniques involved here are not based on the Bethe Ansatz, but instead use a Jordan-Wigner transformation. The first step is to introduce new operators which are fermion operators, unlike the spin operators considered previously. The Jordan-Wigner transformation then involves combining these fermion operators into new 'quasiparticle' operators that are still fermions. The ground state is then a state in which all eigenstates of these operators are occupied up to the 'Fermi surface'. It is then possible to calculate the ground state energy of this system.

6.1 Introduction

This is again a 1D system (i.e. chain) with periodic boundary conditions and nearest-neighbour only interactions. However the form of the interaction is different. The Hamiltonian is

$$\mathcal{H} = J \sum_j \left[(1 + \gamma) S_j^x S_{j+1}^x + (1 - \gamma) S_j^y S_{j+1}^y \right]. \tag{6.1}$$

Again we consider only the case $S = \frac{1}{2}$ since only this value has an exact solution.

The XY model is interesting since the exact solution uses a method which is very different to the Bethe Ansatz. This involves a transformation from spin variables (i.e. angular momentum operators) to fermion operators. The method is called the *Jordan-Wigner transformation* and was first introduced in 1928 [1] for use in early work on second quantisation. It was applied to the XY model in a famous paper by Lieb et al. [2].

Because the procedure is somewhat lengthy it is divided here into sections.

Parkinson, J.B., Farnell, D.J.J.: *The XY Model*. Lect. Notes Phys. **816**, 61–75 (2010)
DOI 10.1007/978-3-642-13290-2_6

6.2 Change from Spin Operators to Fermion Operators

The spin operators for a single site i satisfy

$$S_i^+|-\rangle = |+\rangle \quad S_i^+|+\rangle = 0 \tag{6.2}$$

$$S_i^-|-\rangle = 0 \quad S_i^-|+\rangle = |-\rangle \tag{6.3}$$

$$S_i^z|-\rangle = -\frac{1}{2}|-\rangle \quad S_i^z|+\rangle = \frac{1}{2}|+\rangle \tag{6.4}$$

Therefore

$$(S_i^- S_i^+ + S_i^+ S_i^-)|+\rangle = S_i^+|-\rangle = |+\rangle$$

and

$$(S_i^- S_i^+ + S_i^+ S_i^-)|-\rangle = S_i^-|+\rangle = |-\rangle$$

so clearly

$$S_i^- S_i^+ + S_i^+ S_i^- = 1. \tag{6.5}$$

Also

$$S_i^{-2} = S_i^{+2} = 0. \tag{6.6}$$

These results apply to a single site i. Any 2 spin operators referring to different sites will commute, e.g.

$$[S_i^-, S_j^+] \equiv S_i^- S_j^+ - S_j^+ S_i^- = 0. \quad (i \neq j)$$

Therefore, for $i \neq j$,

$$S_i^- S_j^+ + S_j^+ S_i^- = 2S_i^- S_j^+ \tag{6.7}$$

$$S_i^- S_j^- + S_j^- S_i^- = 2S_i^- S_j^- \tag{6.8}$$

$$S_i^+ S_j^+ + S_j^+ S_i^+ = 2S_i^+ S_j^+. \tag{6.9}$$

Now introduce a pair of *fermion* operators c_i and c_i^+ for each site. Since they are fermions they anticommute, even when referring to different sites,

$$\left\{c_i, c_j^+\right\} \equiv c_i c_j^+ + c_j^+ c_i = \delta_{ij} \tag{6.10}$$

$$\left\{c_i, c_j\right\} = 0 \tag{6.11}$$

$$\left\{c_i^+, c_j^+\right\} = 0 \tag{6.12}$$

For a single site (6.5) and (6.6) would be satisfied by putting $S_i^- = c_i$ and $S_i^+ = c_i^+$. Unfortunately, this is incorrect for different sites as (6.7), (6.8), and (6.9) do not agree with (6.10), (6.11) and (6.12), respectively. For example

$$\left\{ S_i^-, S_j^+ \right\} = 2S_i^- S_j^+ \neq 0. \qquad (i \neq j)$$

Jordan and Wigner showed that the spin operators S_i^- and S_i^+ can be represented *exactly* in terms of the fermions c_i etc by writing

$$
\begin{aligned}
S_1^- &= c_1 & S_1^+ &= c_1^+ \\
S_2^- &= [\exp(i\pi c_1^+ c_1)]c_2 & S_2^+ &= c_2^+ \exp(-i\pi c_1^+ c_1) \\
S_i^- &= Q_i c_i \quad i \geq 1 & S_i^+ &= c_i^+ Q_i^+ \quad i \geq 1
\end{aligned}
\qquad (6.13)
$$

where $Q_i = \exp\left[i\pi \sum_{l=1}^{i-1} c_l^+ c_l \right]$.

To prove that the spin operators written using (6.13) commute when on different sites requires first that we prove a number of relations and introduce new operators n_i and T_i as well as Q_i.

Let $|+\rangle$ and $|-\rangle$ be a basis for the ith site. The fermion operators act on these as follows

$$
\begin{aligned}
c_i^+|+\rangle &= 0 & c_i^+|-\rangle &= |+\rangle \\
c_i|+\rangle &= |-\rangle & c_i|-\rangle &= 0
\end{aligned}
$$

For any given site, if we know effect of an operator on $|+\rangle$ and $|-\rangle$, then we know everything we need about that operator, since any state is a linear combination of $|+\rangle$ and $|-\rangle$. Conversely, if two operators $\hat{A}_i$ and $\hat{B}_i$ have the same effect on $|+\rangle$ and on $|-\rangle$ then they are equal.

Now define the operator

$$n_i \equiv c_i^+ c_i$$

so clearly $n_i^+ = n_i$ and

$$n_i|+\rangle = c_i^+ c_i|+\rangle = c_i^+|-\rangle = |+\rangle \qquad (6.14)$$

and

$$n_i|-\rangle = c_i^+ c_i|-\rangle = 0. \qquad (6.15)$$

We say that n_i is the number operator, which 'counts' the value of the z-component of the angular momentum relative to state $|-\rangle$, since $n_i|-\rangle = 0|-\rangle$ and $n_i|+\rangle = 1|+\rangle$.

Now define

$$T_i \equiv e^{i\pi c_i^+ c_i} = e^{i\pi n_i}.$$

Clearly

$$
\begin{aligned}
T_i|+\rangle &= e^{i\pi n_i}|+\rangle = e^{i\pi 1}|+\rangle = -|+\rangle \\
T_i|-\rangle &= e^{i\pi n_i}|-\rangle = e^0|-\rangle = |-\rangle
\end{aligned}
$$

which tells us all we need to know about T_i.

Similarly $T_i^+ = e^{-i\pi n_i}$ so

$$T_i^+|+\rangle = e^{-i\pi n_i}|+\rangle = e^{-i\pi 1}|+\rangle = -|+\rangle$$
$$T_i^+|-\rangle = e^{-i\pi n_i}|-\rangle = e^0|-\rangle = |-\rangle$$

and since the effect of T_i^+ is the same as that of T_i on each of the two basis states we can put

$$T_i^+ = T_i. \tag{6.16}$$

Also $T_i^2|+\rangle = |+\rangle$ and $T_i^2|-\rangle = |-\rangle$ and therefore we can replace T_i^2 by 1, i.e.

$$T_i^2 = 1. \tag{6.17}$$

Finally note that

$$
\begin{aligned}
n_i n_j &= c_i^+ c_i c_j^+ c_j & i \neq j \\
&= -c_i^+ c_j^+ c_i c_j \\
&= +c_j^+ c_i^+ c_i c_j \\
&= -c_j^+ c_i^+ c_j c_i \\
&= +c_j^+ c_j c_i^+ c_i = n_j n_i \quad (i \neq j)
\end{aligned}
$$

and since this is trivially true for $i = j$ also we have shown that $[n_i, n_j] = 0$ for all i, j. It follows immediately that $[T_i, T_j] = 0$ for all i, j. Note that these are commutators, not anticommutators.

Now $Q_i = \exp\left[i\pi \sum_{l=1}^{i-1} n_l\right]$ and since all n_l commute

$$Q_i = \exp \sum_{l=1}^{i-1}(i\pi n_l) = \prod_{l=1}^{i-1} e^{i\pi n_l} = \prod_{l=1}^{i-1} T_l$$

and

$$Q_i^+ = \exp \sum_{l=1}^{i-1}(-i\pi n_l) = \prod_{l=1}^{i-1} e^{-i\pi n_l} = \prod_{l=1}^{i-1} T_l^+ = \prod_{l=1}^{i-1} T_l = Q_i$$

(recall that $e^{A+B} = e^A e^B$ provided A and B commute.)

Now consider $[c_i, n_j]$ for $i \neq j$

$$c_i n_j = c_i c_j^+ c_j = -c_j^+ c_i c_j$$
$$= +c_j^+ c_j c_i = n_j c_i$$

therefore $[c_i, n_j] = 0$ for $i \neq j$ and likewise $[c_i, T_j] = [c_i^+, T_j] = 0$ for $i \neq j$.
However

$$c_i T_i |+\rangle = -c_i |+\rangle = -|-\rangle \qquad (6.18)$$
$$T_i c_i |+\rangle = T_i |-\rangle = |-\rangle \qquad (6.19)$$
$$c_i T_i |-\rangle = c_i |-\rangle = 0 \qquad (6.20)$$
$$T_i c_i |-\rangle = 0 \qquad (6.21)$$

therefore

$$(c_i T_i + T_i c_i)|+\rangle = 0 \qquad\qquad (c_i T_i - T_i c_i)|+\rangle = -2|-\rangle$$
$$(c_i T_i + T_i c_i)|-\rangle = 0 \qquad\qquad (c_i T_i - T_i c_i)|-\rangle = 0$$

so $\{c_i, T_i\} = 0$ but $[c_i, T_i] \neq 0$.
Similarly $\{c_i^+, T_i\} = 0$ but $[c_i^+, T_i] \neq 0$.
Also

$$[c_i, Q_i] = \left[c_i, \prod_{l=1}^{i-1} T_l \right] = 0 \quad \text{since } l \neq i$$

and similarly $[c_i^+, Q_i] = 0$.

Finally we need to show that spin operators on different sites commute when written in terms of the fermion operators as given in Eq. (6.13). We shall give the proof for one case only, namely $[S_i^-, S_j^+]$ where $j > i$.

$$S_i^- S_j^+ = Q_i c_i c_j^+ Q_j^+ = c_i Q_i Q_j c_j^+$$
$$= c_i \prod_{l=1}^{i-1} T_l \prod_{m=1}^{j-1} T_m c_j^+$$
$$= c_i \prod_{l=i}^{j-1} T_l c_j^+ \quad \text{since } T_n^2 = 1$$
$$= c_i c_j^+ \prod_{l=i}^{j-1} T_l \quad \text{since } l \neq j$$

Similarly

$$S_j^+ S_i^- = c_j^\dagger Q_j^\dagger Q_i c_i = c_j^\dagger Q_j Q_i c_i$$

$$= c_j^\dagger \prod_{l=i}^{j-1} T_l\, c_i$$

$$= c_j^\dagger T_i c_i \prod_{l=i+1}^{j-1} T_l$$

$$= -c_j^\dagger c_i T_i \prod_{l=i+1}^{j-1} T_l$$

$$= c_i c_j^\dagger \prod_{l=i}^{j-1} T_l = S_i^- S_j^+$$

Hence $[S_i^-, S_j^+] = 0$.

Other cases are similar.

Although this transformation looks complicated, the result on the Hamiltonian (6.1) is rather simple. First we rewrite (6.1)

$$S_j^x S_{j+1}^x = \frac{1}{2}(S_j^+ + S_j^-)\frac{1}{2}(S_{j+1}^+ + S_{j+1}^-)$$

$$= \frac{1}{4}(S_j^+ S_{j+1}^+ + S_j^- S_{j+1}^-) + \frac{1}{4}(S_j^- S_{j+1}^+ + S_j^+ S_{j+1}^-)$$

$$S_j^y S_{j+1}^y = \frac{1}{2i}(S_j^+ - S_j^-)\frac{1}{2i}(S_{j+1}^+ - S_{j+1}^-)$$

$$= -\frac{1}{4}(S_j^+ S_{j+1}^+ + S_j^- S_{j+1}^-) + \frac{1}{4}(S_j^- S_{j+1}^+ + S_j^+ S_{j+1}^-)$$

Thus

$$\mathcal{H} = J \sum_{j=1}^N \left[(1+\gamma)S_j^x S_{j+1}^x + (1-\gamma)S_j^y S_{j+1}^y \right]$$

$$= \frac{J}{2} \sum_{j=1}^N \left[(S_j^- S_{j+1}^+ + S_j^+ S_{j+1}^-) + \gamma(S_j^+ S_{j+1}^+ + S_j^- S_{j+1}^-) \right] \qquad (6.22)$$

This form shows why the treatment of the XY-model is so different to that of the isotropic Heisenberg model. The terms with γ in (6.22) connect states with different S_T^z so this is clearly not a constant of the motion, i.e. the eigenstates are not combinations of basis states which all have the same S_T^z. The only exception to this is, of course, the case $\gamma = 0$, which is sometimes known as the 'XX-model'.

Now change to fermion operators

$$S_j^- S_{j+1}^+ = Q_j c_j c_{j+1}^+ Q_{j+1}$$
$$= c_j T_j c_{j+1}^+$$

But from (6.18) and (6.20) we have $c_j T_j |+\rangle = -|-\rangle$ and $c_j T_j |-\rangle = 0.$

Comparing this with $c_j |+\rangle = |-\rangle$ and $c_j |-\rangle = 0$ we see that $c_j T_j = -c_j$ so $S_j^- S_{j+1}^+ = -c_j c_{j+1}^+ = c_{j+1}^+ c_j$. A similar treatment of the other three terms in Eq. (6.22) leads to the following set

$$S_j^- S_{j+1}^+ = c_{j+1}^+ c_j \tag{6.23}$$
$$S_j^+ S_{j+1}^- = c_j^+ c_{j+1} \tag{6.24}$$
$$S_j^+ S_{j+1}^+ = c_j^+ c_{j+1}^+ \tag{6.25}$$
$$S_j^- S_{j+1}^- = c_{j+1} c_j \tag{6.26}$$

(Note the ordering of these operators).

There is a problem at the end however. For $j = N$ we get terms of the form

$$S_N^+ S_1^- = Q_N c_N^+ c_1 \neq c_N^+ c_1$$

This occurs because the transformation to fermion operators Eq. (6.13), involves a phase factor Q_i which does not satisfy periodic boundary conditions. Thus

$$\mathcal{H} = \frac{J}{2} \sum_{j=1}^{N} [(c_{j+1}^+ c_j + c_j^+ c_{j+1}) + \gamma(c_j^+ c_{j+1}^+ + c_{j+1} c_j)]$$
$$- \frac{J}{2} [(c_1^+ c_N + c_N^+ c_1) + \gamma(c_N^+ c_1^+ + c_1 c_N)]$$
$$\frac{J}{2} Q_N [(c_N c_1^+ + c_N^+ c_1) + \gamma(c_N^+ c_1^+ + c_N c_1)]$$

Because the last two terms do not involve a sum over sites, we expect that in the limit $N \to \infty$ they can be neglected. In fact, Lieb et al. [2] show how to treat them correctly and their result confirms that it is valid to neglect them in this limit. Thus we can write

$$\mathcal{H} = \frac{J}{2} \sum_{j=1}^{N} [(c_{j+1}^+ c_j + c_j^+ c_{j+1}) + \gamma(c_j^+ c_{j+1}^+ + c_{j+1} c_j)]. \tag{6.27}$$

Again note that the ordering of the operators is important here.

6.3 Fourier Transform

Equation (6.27) is a quadratic Hamiltonian involving only fermion operators. The first step in diagonalising it is to make use of the translational invariance by introducing Fourier transformed operators d_k and d_k^+

$$d_k = \frac{1}{\sqrt{N}} \sum_{j=1}^{N} e^{-ikj} c_j \tag{6.28}$$

$$d_k^+ = \frac{1}{\sqrt{N}} \sum_{j=1}^{N} e^{ikj} c_j^+ \tag{6.29}$$

with $k = \lambda \frac{2\pi}{N}$ where λ is an integer such that $\lambda = \left(-\frac{N}{2} + 1\right), \ldots \left(\frac{N}{2}\right)$. Clearly $-\pi < k \leq \pi$.

The reverse transform is

$$c_j = \frac{1}{\sqrt{N}} \sum_{k} e^{ikj} d_k \tag{6.30}$$

$$c_j^+ = \frac{1}{\sqrt{N}} \sum_{k} e^{-ikj} d_k^+ . \tag{6.31}$$

Using the properties of the c's we can easily show that the d's are fermions also, i.e.

$$\{d_{k_1}, d_{k_2}\} = \left\{d_{k_1}^+, d_{k_2}^+\right\} = 0$$

$$\left\{d_{k_1}, d_{k_2}^+\right\} = \delta_{k_1 k_2}$$

(In showing these we need to use $\sum_{j} e^{i(k_1 - k_2)j} = N\delta_{k_1 k_2}$, etc.)

Now rewrite each term in $\mathcal{H}$ in terms of these:

$$\sum_{j} c_{j+1}^+ c_j = \sum_{j} \frac{1}{N} \sum_{k_1} \sum_{k_2} e^{-ik_1(j+1)} e^{ik_2 j} d_{k_1}^+ d_{k_2}$$

$$= \frac{1}{N} \sum_{k_1} \sum_{k_2} e^{-ik_1} N\delta_{k_1 k_2} d_{k_1}^+ d_{k_2}$$

$$= \sum_{k} e^{-ik} d_k^+ d_k ,$$

and similarly

$$\sum_j c_j^+ c_{j+1} = \sum_k e^{ik} d_k^+ d_k,$$

$$\sum_j c_j^+ c_{j+1}^+ = \sum_k e^{ik} d_k^+ d_{-k}^+,$$

$$\sum_j c_{j+1} c_j = \sum_k e^{ik} d_k d_{-k}.$$

so

$$\mathcal{H} = J \sum_k \left[\cos k (d_k^+ d_k) + \gamma \frac{e^{ik}}{2} (d_k^+ d_{-k}^+ + d_k d_{-k}) \right] \qquad (6.32)$$

This use of the translational invariance has partially diagonalised the Hamiltonian. There is now no coupling between states with different $|k|$. It remains to diagonalise the coupled k and $-k$ terms. This is done by introducing new fermion operators which are linear combinations of k and $-k$ operators. A linear combination of this kind is called a *Bogoliubov* transformation [3], although a similar transformation was done earlier by Holstein and Primakoff [4] in connection with spin waves in a ferromagnet with dipole–dipole interactions and a magnetic field as well as exchange interactions. We shall also refer to these operators as *quasiparticle* operators.

6.4 Quasiparticle Operators

Clearly operators d_k and d_{-k} are linked in (6.32) so instead of summing $-\pi < k < \pi$ we combine the k and $-k$ terms and sum $0 \le k < \pi$

$$\mathcal{H} = J \sum_{k=0}^{\pi} \left[\cos k (d_k^+ d_k + d_{-k}^+ d_{-k}) \right.$$

$$\left. + \gamma \left\{ \frac{e^{ik}}{2} (d_k^+ d_{-k}^+ + d_k d_{-k}) + \frac{e^{-ik}}{2} (d_{-k}^+ d_k^+ + d_{-k} d_k) \right\} \right]$$

and since $d_{-k}^+ d_k^+ = -d_k^+ d_{-k}^+$ and $d_{-k} d_k = -d_k d_{-k}$

$$\mathcal{H} = J \sum_{k=0}^{\pi} \left[\cos k (d_k^+ d_k + d_{-k}^+ d_{-k}) + \gamma \{ i \sin k (d_k^+ d_{-k}^+ + d_k d_{-k}) \} \right] \qquad (6.33)$$

There are basically two fermion operators in this expression, namely d_k and d_{-k} together with their Hermitian adjoints. To diagonalise it we look for two different linear combinations, η_k and η_{-k}, of d_k and d_{-k} which are also fermions and such that the Hamiltonian has the form

$$\mathcal{H} = \sum_{k=0}^{\pi} \left[\Lambda_{1k} \eta_k^+ \eta_k + \Lambda_{2k} \eta_{-k}^+ \eta_{-k} + X_k \right]. \tag{6.34}$$

where X_k is a constant.

In fact the linear combinations we require are not between d_k and d_{-k} but between d_k and d_{-k}^+. The first one is

$$\eta_k = A_k d_k + B_k d_{-k}^+. \tag{6.35}$$

For the second one, instead of a linear combination of d_k and d_{-k}^+, it is convenient to take a linear combination of d_k^+ and d_{-k}

$$\eta_{-k} = C_k d_{-k} + D_k d_k^+. \tag{6.36}$$

Since η_{-k} and η_{-k}^+ occur together in (6.34) this is permissible.

In order for these η_k to be also fermion operators, i.e. for the transformation to be canonical, we require that

$$\{\eta_k, \eta_k\} = \{\eta_k^+, \eta_k^+\} = 0 \tag{6.37}$$

$$\{\eta_{-k}, \eta_{-k}\} = \{\eta_{-k}^+, \eta_{-k}^+\} = 0 \tag{6.38}$$

$$\{\eta_k, \eta_{-k}^+\} = \{\eta_{-k}, \eta_k^+\} = 0 \tag{6.39}$$

$$\{\eta_k, \eta_k^+\} = \{\eta_{-k}, \eta_{-k}^+\} = 1 \tag{6.40}$$

$$\{\eta_k, \eta_{-k}\} = \{\eta_k^+, \eta_{-k}^+\} = 0 \tag{6.41}$$

(6.37) and (6.38) are automatically satisfied, e.g.

$$\{\eta_k, \eta_k\} = 2\eta_k^2 = 2(A_k d_k + B_k d_{-k}^+)(A_k d_k + B_k d_{-k}^+)$$
$$= 2\{A_k^2 d_k^2 + A_k B_k (d_k d_{-k}^+ + d_{-k}^+ d_k) + B_k^2 (d_{-k}^+)^2\}$$
$$= 0$$

using $d_k^2 = (d_{-k}^+)^2 = 0$ and $\{d_k, d_{-k}^+\} = 0$.
(6.39) is also automatically satisfied, e.g.

$$\{\eta_k, \eta_{-k}^+\} = (A_k d_k + B_k d_{-k}^+)(C_k^* d_{-k}^+ + D_k^* d_k) + (C_k^* d_{-k}^+ + D_k^* d_k)(A_k d_k + B_k d_{-k}^+)$$
$$= A_k C_k^* (d_k d_{-k}^+ + d_{-k}^+ d_k) + A_k D_k^* (d_k d_k + d_k d_k)$$
$$+ B_k C_k^* (d_{-k}^+ d_{-k}^+ + d_{-k}^+ d_{-k}^+) + B_k D_k^* (d_{-k}^+ d_k + d_k d_{-k}^+) = 0$$

since all the brackets in the second line are anticommutators which equal zero.

Equation (6.40) requires

$$
\begin{aligned}
\{\eta_k, \eta_k^+\} &= (A_k d_k + B_k d_{-k}^+)(A_k^* d_k^+ + B_k^* d_{-k}) + (A_k^* d_k^+ + B_k^* d_{-k})(A_k d_k + B_k d_{-k}^+) \\
&= |A_k|^2 (d_k d_k^+ + d_k^+ d_k) + A_k B_k^* (d_k d_{-k} + d_{-k} d_k) \\
&+ B_k A_k^* (d_{-k}^+ d_k^+ + d_k^+ d_{-k}^+) + |B_k|^2 (d_{-k}^+ d_{-k} + d_{-k} d_{-k}^+) \\
&= |A_k|^2 + |B_k|^2 = 1
\end{aligned} \tag{6.42}
$$

and similarly

$$
|C_k|^2 + |D_k|^2 = 1 \tag{6.43}
$$

Finally (6.41) gives

$$
\begin{aligned}
\{\eta_k, \eta_{-k}\} &= (A_k d_k + B_k d_{-k}^+)(C_k d_{-k} + D_k d_k^+) + (C_k d_{-k} + D_k d_k^+)(A_k d_k + B_k d_{-k}^+) \\
&= A_k C_k (d_k d_{-k} + d_{-k} d_k) + A_k D_k (d_k d_k^+ + d_k^+ d_k) \\
&+ B_k C_k (d_{-k}^+ d_{-k} + d_{-k} d_{-k}^+) + B_k D_k (d_{-k}^+ d_k^+ + d_k^+ d_{-k}^+) \\
&= A_k D_k + B_k C_k = 0
\end{aligned} \tag{6.44}
$$

together with its complex conjugate.

6.5 Quasiparticle Energies

With the Hamiltonian now written in the form (6.34) then the corresponding quasi-particle energies are Λ_{1k} for η_k and Λ_{2k} for η_{-k}, the first of which satisfies

$$
[\eta_k, \mathcal{H}] = \Lambda_{1k} \eta_k. \tag{6.45}
$$

Note the commutator in this equation even though the η_k are fermions like the d_k.
  The d_k satisfy the following commutation relations:

$$
[d_k, d_{k_1}^+ d_{k_1}] = \delta_{kk_1} d_k \tag{6.46}
$$
$$
[d_k, d_{k_1}^+ d_{-k_1}^+] = (\delta_{kk_1} - \delta_{k,-k_1}) d_{-k}^+ \tag{6.47}
$$
$$
[d_k, d_{k_1} d_{-k_1}] = 0 \tag{6.48}
$$

Hence

$$
\begin{aligned}
[d_k, \mathcal{H}] &= J \sum_{k_1} \cos k \, \delta_{kk_1} \, d_k + \frac{J\gamma}{2} \sum_{k_1} e^{ik} \left(\delta_{kk_1} - \delta_{k,-k_1} \right) d_{-k}^+ \\
&= J \cos k \, d_k + J i \gamma \sin k \, d_{-k}^+
\end{aligned} \tag{6.49}
$$

This shows clearly the coupling between d_k and d_{-k}^+.

For d_{-k}^+ the commutation relations are

$$\left[d_{-k}^+, d_{k_1}^+ d_{k_1}\right] = -\delta_{k,-k_1} d_{-k}^+ \tag{6.50}$$

$$\left[d_{-k}^+, d_{k_1}^+ d_{-k_1}^+\right] = 0 \tag{6.51}$$

$$\left[d_{-k}^+, d_{k_1} d_{-k_1}\right] = \left(\delta_{k,-k_1} - \delta_{kk_1}\right) d_k \tag{6.52}$$

leading to

$$[d_{-k}^+, \mathcal{H}] = J \sum_{k_1} \cos k_1 (-\delta_{k,-k_1}) d_{-k}^+ + \frac{J\gamma}{2} \sum_{k_1} e^{ik_1} (\delta_{k,-k_1} - \delta_{kk_1}) d_k$$

$$= -J \cos k \, d_{-k}^+ - Ji\gamma \sin k \, d_k. \tag{6.53}$$

Using (6.49) and (6.53) gives

$$[\eta_k, \mathcal{H}] = A_k (J \cos k d_k + Ji\gamma \sin k d_{-k}^+) + B_k (-J \cos k d_{-k}^+ - Ji\gamma \sin k d_k) \tag{6.54}$$

and from (6.45)

$$[\eta_k, \mathcal{H}] = \Lambda_{1k} (A_k d_k + B_k d_{-k}^+). \tag{6.55}$$

Comparing the coefficients of d_k and d_{-k}^+ in these two equations we obtain simultaneous equations for A_k and B_k

$$\Lambda_{1k} A_k = A_k J \cos k - B_k Ji\gamma \sin k \tag{6.56}$$

$$\Lambda_{1k} B_k = A_k Ji\gamma \sin k - B_k J \cos k \tag{6.57}$$

which have a non trivial solution if

$$\begin{vmatrix} \Lambda_{1k} - J \cos k & Ji\gamma \sin k \\ -Ji\gamma \sin k & \Lambda_k + J \cos k \end{vmatrix} = 0$$

$$\Lambda_{1k}^2 - J^2 \cos^2 k - J^2 \gamma^2 \sin^2 k = 0$$

$$\Lambda_{1k}^2 = J^2 (\cos^2 k + \gamma^2 \sin^2 k) \tag{6.58}$$

i.e.

$$\Lambda_{1k} = \pm J L_k \quad \text{where } L_k = +\sqrt{(\cos^2 k + \gamma^2 \sin^2 k)}. \tag{6.59}$$

A Plot L_k vs k is shown in Fig. 6.1 for three values of γ.

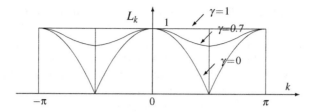

Fig. 6.1 Plot of $L_k = \Lambda_{1k}/J$ vs. k for three different values of the anisotropy γ

6.6 Ground State Energy of the XY-Model

The fermion operators corresponding to these eigenvalues have the form

$$\eta_k = A_k d_k + B_k d^+_{-k} \tag{6.60}$$

with

$$\eta^+_k = A^*_k d^+_k + B^*_k d_{-k} \tag{6.61}$$

From (6.57) and (6.58)

$$B_k = \frac{Ji\gamma \sin k}{\Lambda_{1k} + J \cos k} A_k \tag{6.62}$$

Clearly the ratio $\dfrac{B_k}{A_k}$ is different for the two choices of Λ_{1k}.

The corresponding solution for $k \to -k$, obtained from $[\eta_{-k}, \mathcal{H}]$, gives $\Lambda_{2k} = \pm JL_k$, the same as Λ_{1k}, and two corresponding solutions for η_{-k}.

Putting

$$\eta_{-k} = C_k d_{-k} + D_k d^+_k \tag{6.63}$$

gives

$$D_k = \frac{-Ji\gamma \sin k}{\Lambda_{2k} + J \cos k} C_k \tag{6.64}$$

and again the ratio $\dfrac{D_k}{C_k}$ is different for the two choices of Λ_{2k}.

Note that η^+_{-k} has same form as η_k but with $A_k \to D^*_k$ and $B_k \to C^*_k$. This means that there are not four different terms in the Hamiltonian corresponding to two different η_k and two different η_{-k} but only two linearly independent terms. The choice of two linearly independent terms from the four solutions is arbitrary.

The different choices give Hamiltonians which have a similar form but differ by an additive constant.

The conventional choice is to take $\Lambda_{2k} = \Lambda_{1k} = +JL_k$, together with the corresponding choice of ratios $\frac{B_k}{A_k}$ and $\frac{D_k}{C_k}$ from (6.62) and (6.64).

This choice has the result that the excitations from the ground state have positive energy so that the ground state is the state with zero excitations. If $|\phi\rangle$ is the ground state, whose detailed form is not given here, then

$$\eta_k^+ \eta_k |\phi\rangle = \eta_{-k}^+ \eta_{-k} |\phi\rangle = 0 \quad \text{for all } k.$$

Using Eq. (6.34) the total energy of the ground state is

$$E_G = \sum_{k=0}^{\pi} X_k \tag{6.65}$$

and with the X_k determined by the condition that (6.34) and (6.33) are equal:

$$E_G = -J \sum_{k=0}^{\pi} L_k(|D_k|^2 + |B_k|^2).$$

Using Eqs. (6.42), (6.43), (6.62), and (6.64) this becomes

$$= -J \sum_{k=0}^{\pi} (L_k - \cos k)$$

$$= -J \sum_{k=0}^{\pi} L_k$$

In deriving this result we have also used $(L_k + \cos k)^2 + \gamma^2 \sin^2 k = 2L_k(L_k + \cos k)$.

Changing to an integral by replacing $\sum_{k=0}^{\pi} = \frac{N}{2\pi} \int_0^{\pi} dk$

$$\frac{E_G}{N} = -\frac{J}{2\pi} \int_0^{\pi} dk \left[\cos^2 k + \gamma^2 \sin^2 k\right]^{\frac{1}{2}}$$

which is an elliptic integral. Note that the positive root is required here.

Important special cases are

$$\gamma = 0 \quad \text{'XX model'} \quad \frac{E_G}{NJ} = -\frac{1}{\pi},$$

$$\gamma = 1 \quad \text{Ising case} \quad \frac{E_G}{NJ} = -\frac{1}{2}.$$

Note that the ground state $|\phi\rangle$ is a state for which $\eta_k|\phi\rangle = 0$ for all k. However the η_k are obviously not destruction operators with respect to the aligned state $|0\rangle$ since if they were then the aligned state would be the ground state. Another way of saying this is that the fermion operators η_k are the destruction operators of 'quasiparticles' unlike the original fermi operators c_k which correspond to spin reversals. In fact $|\phi\rangle$ is a linear combination of states with any even number from 0 to N of reversed spins with respect to $|0\rangle$. Only even numbers occur because the Hamiltonian Eq. (6.22) only connects states differing by two reversals. On average, however, $\frac{N}{2}$ spins will be reversed at any one time.

Some of the reasons why LSM's result is so important are as follows:

a. It gives an explicit simple form for the ground state energy.
b. Using the properties of the η_k, LSM were able to calculate the *correlations* for the XY-model. They showed that for $\gamma = 0$ (isotropic case)

$$\langle S_i^z S_{i+R}^z \rangle \xrightarrow[R \to \infty]{} \frac{1}{\pi^2 R^2}$$

$$\langle S_i^x S_{i+R}^x \rangle = \langle S_i^y S_{i+R}^y \rangle \xrightarrow[R \to \infty]{} \frac{const.}{R^{\frac{1}{2}}}$$

so that there is *no long range order* for $\gamma = 0$.

The case $\gamma \neq 0$ is more complicated but they showed that there *is* long range order which they were able to calculate.
c. The thermodynamics, i.e. non-zero temperature, properties can also be calculated since the system consists of non-interacting fermi quasiparticles. In particular LSM were able to calculate the free energy per spin as a function of T.
d. Correlation functions in the presence of a magnetic field have also been calculated [5, 6].

References

1. Jordan, P., Wigner, E.: Z. Phys. **47**, 631 (1928)
2. Lieb, E., Schultz, T., Mattis, D.: Ann. Phys. **16**, 407 (1961)
3. Bogoliubov, N.N.: Physikalische Abhandlungen der Sowjetunion **6**, 1, 113, 229 (1962)
4. Holstein, T., Primakof, H.: Phys. Rev. **58**, 1098 (1940)
5. Niemeijer, Th.: Physica **36**, 377–419 (1967)
6. Niemeijer, Th.: Physica **39**, 313–326 (1968)

Chapter 7
Spin-Wave Theory

Abstract All of the techniques described in the previous chapters are exact methods which work only for particular, but important, special cases. In this chapter we describe a much more general method, due originally to Anderson, which is widely used to obtain results for many different systems. It is referred to as 'spin-wave theory' as it gives results for the energies of the elementary excitations or spin waves. The method works for both ferromagnets and antiferromagnets, it works in 2D and 3D as well as 1D, and it works for arbitrary spin – not just for spin-1/2. The ferromagnetic version is rather simple and the results are usually exact. The antiferromagnetic version is more complicated and the results are approximate. However, these results are still in reasonable correspondence with exact results, where these are known, and the best of other approximate methods.

7.1 Introduction

The special mathematical techniques introduced for $S = \frac{1}{2}$ chains, namely the Bethe Ansatz for Heisenberg coupling and the Jordan-Wigner transformation for the XY-model, do not work for higher S or higher dimensions. Spin-wave theory is a much more general method of studying spin-models introduced by Anderson (1952) [1] for ferromagnets and later applied by Oguchi (1963) [2] to antiferromagnets.

Spin-wave (SW) theory is not (in general) exact. However, it has many useful features, some of which are

a. It is rather simple.
b. It works for many Hamiltonians and in any number of dimensions.
c. It gives approximate results for non-zero temperature.
d. It works for arbitrary S.
e. It gives a good physical picture of the excitations.

The basic idea of spin-wave theory is to replace the spin operators by bosons. As we have seen, spin operators behave like fermions on a given site, but like bosons where different sites are concerned. Hence we need to find a way of handling the spin operators on a given site in terms of bosons.

Parkinson, J.B., Farnell, D.J.J.: *Spin-Wave Theory*. Lect. Notes Phys. **816**, 77–88 (2010)
DOI 10.1007/978-3-642-13290-2_7

Recall that for $S = \frac{1}{2}$ and a single site

$$S^+|-\rangle = |+\rangle \qquad\qquad\qquad S^+|+\rangle = 0$$

$$S^-|-\rangle = 0 \qquad\qquad\qquad S^-|+\rangle = |-\rangle$$

$$S^z|-\rangle = -\frac{1}{2}|-\rangle \qquad\qquad S^z|+\rangle = \frac{1}{2}|+\rangle$$

and all other spin operators can be written in terms of these.

For a general spin S and a single site the corresponding basis is again the eigenstates of S^z written as $|m\rangle$ where the $2S + 1$ values of m are

$$m = -S, -S + 1, \ldots, S$$

and the states $|m\rangle$ are orthogonal and normalised (i.e. orthonormal). We refer to $S - m$ as the number of deviations from the state $|S\rangle$, the state of maximum S^z.

The corresponding operators and eigenvalues are

$$S^+|m\rangle = \sqrt{(S - m)(S + m + 1)}|m + 1\rangle \qquad (7.1)$$

$$S^-|m\rangle = \sqrt{(S - m + 1)(S + m)}|m - 1\rangle \qquad (7.2)$$

$$S^z|m\rangle = m|m\rangle. \qquad (7.3)$$

Special cases for $m = \pm S$ are $S^+|S\rangle = 0$ and $S^-|-S\rangle = 0$.

We now introduce *boson* operators for a single site which reproduce most of the above properties. There are various ways of doing this but we shall only consider the most useful and widely used, called the *Holstein-Primakoff* transformation.

Let a^+, a be boson creation and destruction operators. We now interpret the number of bosons as the number of deviations from the state $|S\rangle$. State $|m\rangle$ has $(S - m)$ deviations and the number operator for bosons is $\hat{n} \equiv a^+ a$ so

$$\hat{n}|m\rangle = a^+ a|m\rangle = (S - m)|m\rangle. \qquad (7.4)$$

Clearly we can represent the operator S^z as

$$S^z = S - \hat{n} \qquad (7.5)$$

since $S^z|m\rangle = (S - \hat{n})|m\rangle = [S - (S - m)]|m\rangle = m|m\rangle$. Note that (7.3) is now satisfied.

We now have to represent the S^+ and S^- operators in terms of the bosons. This is done as follows:

$$S^+ = (2S)^{\frac{1}{2}}\sqrt{1 - \frac{\hat{n}}{2S}}\, a \qquad (7.6)$$

$$S^- = a^+(2S)^{\frac{1}{2}}\sqrt{1 - \frac{\hat{n}}{2S}} \qquad (7.7)$$

Proof Clearly $a|m\rangle$ is a state with one less deviation, i.e.

$$a\,|m\rangle \;=\; A_m|m+1\rangle$$

(recall that less deviations mean higher S^Z).

The Hermitian conjugate of this is

$$\langle m|\,a^+ \;=\; A_m^*\,\langle m+1|$$

and the inner product of these gives

$$\langle m|\,a^+a\,|m\rangle \;=\; |A_m|^2\langle m+1|m+1\rangle$$

and therefore $(S-m)\langle m|m\rangle \;=\; |A_m|^2\langle m+1|m+1\rangle.$

Since the states $|m\rangle$ are othonormal, $\langle m|m\rangle \;=\; \langle m+1|m+1\rangle \;=\; 1$, and so, choosing A_m to be real, $A_m = \sqrt{S-m}$.

Similarly

$$a^+|m\rangle \;=\; B_m|m-1\rangle$$

with $B_m = \sqrt{S-m}$.

We can now show that choices (7.6) and (7.7) satisfy (7.1) and (7.2). Using (7.6)

$$
\begin{aligned}
S^+|m\rangle &= (2S)^{\frac{1}{2}}\sqrt{1 - \frac{\hat{n}}{2S}}\; a|m\rangle \\
&= (2S)^{\frac{1}{2}}\sqrt{S-m}\sqrt{1 - \frac{\hat{n}}{2S}}\,|m+1\rangle \\
&= (2S)^{\frac{1}{2}}\sqrt{S-m}\sqrt{1 - \frac{(S-m-1)}{2S}}\,|m+1\rangle \\
&= \sqrt{S-m}\sqrt{2S-S+m+1}\,|m+1\rangle \\
&= \sqrt{(S-m)(S+m+1)}\,|m+1\rangle
\end{aligned}
$$

which agrees with (7.1). By a similar argument (7.2) is satisfied by (7.7).

Note that the Holstein-Primakoff transformation [(7.5), (7.6), and (7.7)] is exact as far as the states $|S\rangle, |S-1\rangle \ldots |-S\rangle$ are concerned. However, in principle it is possible to have more than S bosons, i.e. a state of the form

$$|\psi\rangle \;=\; (a^+)^k|S\rangle \quad \text{where } k > m.$$

These states are unphysical and they can never be reached if we use the exact transformation. However, we shall now approximate the transformation and this allows coupling to the unphysical states. The approximation will only be valid provided the admixture of the unphysical states is 'small' in some sense.

Mathematically it is very difficult to handle a transformation involving square roots. The approximation we shall use is based on the assumption that the states of interest all have small probabilities of having deviations on any particular site and a negligible probability of having two or more deviations on the same site. This is equivalent to saying

a. The total number of deviations N_D is $\ll N$, i.e. $\langle \hat{n} \rangle \ll 1$.
b. Bound states in which deviations cluster together cannot be treated accurately.

With this assumption we approximate the square root as either

$$\sqrt{1 - \frac{\hat{n}}{2S}} \approx 1 \text{ 'simple SW theory'}$$

or

$$\sqrt{1 - \frac{\hat{n}}{2S}} \approx 1 - \frac{\hat{n}}{4S} \text{ 'interacting SW theory'}$$

Interacting spin-wave theory (which is not covered in this book) can deal with small perturbations to spin-waves obtained using simple spin-wave theory, but it cannot deal with bound states.

For simple SW theory we now obtain a very simple result

$$\begin{aligned}
S^+ &\approx (2S)^{\frac{1}{2}} a \\
S^- &\approx (2S)^{\frac{1}{2}} a^+ \\
S^z &= S - a^+ a
\end{aligned} \tag{7.8}$$

with the usual boson commutation relation $[a, a^+] = 1$. Note that (7.8) refers to a single site. For different sites all operators commute so that

$$[a_i, a_j] = [a_i^+, a_j^+] = 0$$
$$[a_i, a_j^+] = \delta_{ij}$$

7.2 Ferromagnetic Spin-Wave Theory

Consider first the Heisenberg model with nearest neighbour ferromagnetic coupling

$$\mathcal{H} = \frac{J}{2} \sum_j \sum_\rho \mathbf{S}_j \cdot \mathbf{S}_{j+\rho}$$

where $J < 0$, j runs over all sites, ρ runs over all ν neighbours. The ground state (both classical and quantum mechanical) will be a state with all atoms aligned. Usually we take this to be the state in which all atoms are in the $|+S\rangle$ state. (Other degenerate ground states can be easily constructed from this state by using the lowering operator for the whole system $\sum_i S_i^-$). Writing

$$\mathbf{S}_j.\mathbf{S}_{j+\rho} = S_j^z S_{j+\rho}^z + \frac{1}{2}(S_j^+ S_{j+\rho}^- + S_j^- S_{j+\rho}^+)$$

and using (7.8) we get

$$
\begin{aligned}
\mathcal{H} \approx & \frac{JN\nu}{2}S^2 - \frac{JS}{2}\sum_j\sum_\rho(a_j^+ a_j + a_{j+\rho}^+ a_{j+\rho}) \\
& + \frac{JS}{2}\sum_j\sum_\rho(a_j^+ a_{j+\rho} + a_{j+\rho}^+ a_j) \\
& + \frac{J}{2}\sum_j\sum_\rho a_j^+ a_j a_{j+\rho}^+ a_{j+\rho}.
\end{aligned}
$$

The last term here involves four boson operators. For consistency with our previous approximation we must neglect this term.

The first term is the energy of the ground state (all N spins up) $|SSSS....S\rangle$.

Put $E_F = \dfrac{JN\nu S^2}{2}$ (recall that for the ferromagnet J is negative here), so that

$$\mathcal{H} \approx E_F - \frac{JS}{2}\sum_j\sum_\rho[a_j^+ a_j + a_{j+\rho}^+ a_{j+\rho} - a_j^+ a_{j+\rho} - a_{j+\rho}^+ a].$$

This quadratic (or bilinear) Hamiltonian is very easy to diagonalise using a Fourier transform. Define new boson operators

$$\alpha_k = \frac{1}{\sqrt{N}}\sum_j e^{ikj}a_j$$

$$\alpha_k^+ = \frac{1}{\sqrt{N}}\sum_j e^{-ikj}a_j^+$$

$$a_j = \frac{1}{\sqrt{N}}\sum_k e^{-ikj}\alpha_k$$

$$a_j^+ = \frac{1}{\sqrt{N}}\sum_k e^{ikj}\alpha_k^+$$

with $[\alpha_{k_1}, \alpha_{k_2}^+] = \delta_{k_1 k_2}$.

In 3D k is a vector. For example for the simple cubic lattice,

$$k = (\lambda_x, \lambda_y, \lambda_z)\frac{2\pi}{n},$$

with $\lambda_{x,y,z} = 0, 1, 2, \ldots, n-1$, where $n = N^{\frac{1}{3}}$. (Assuming that the number of atoms in each direction n is the same for each of the three perpendicular directions)

Using this

$$\mathcal{H} = E_F - \frac{JS}{2} \sum_k \sum_\rho (1 + 1 - e^{-ik\rho} - e^{ik\rho}) \alpha_k^+ \alpha_k$$

$$= E_F + \sum_k \varepsilon_k \alpha_k^+ \alpha_k$$

where

$$\varepsilon_k = -JS \sum_\rho (1 - \cos k\rho)$$

These are the energies, relative to the fully aligned ground state, of the ferromagnetic spin-waves.

In 1D k is a scalar given by $k = \lambda \frac{2\pi}{N}$, with $\lambda = 0, 1, 2, \ldots N - 1$. The energy is

$$\varepsilon_k = -JS[2 - \cos k - \cos(-k)]$$
$$= -2JS(1 - \cos k)$$

so for $S = \frac{1}{2}$

$$\varepsilon_k = -J(1 - \cos k)$$

which are precisely the energies obtained earlier for the 1-deviation states. The difference now is that we can excite any number of bosons and they will always have energy ε_k. Previously, when 2-deviation states were treated exactly we found both 'free' spin-waves and bound states. In this approximation the corrections to the 'free' state are omitted, and the bound states are not obtained at all.

7.3 Antiferromagnetic Spin-Wave Theory

The SW approximation is in some ways more interesting when applied to antiferromagnets. Classically these tend to align with neighbouring atoms antiparallel. We shall consider only bipartite lattices which can be divided into two sublattices, such that all the nearest neighbours of any atom lie on the opposite sublattice, e.g. chain, square, honeycomb, simple cubic, b.c.c., etc. (Non-bipartite lattices are normally frustrated, and the study of such lattices is much more complex and not considered here.) For these lattices the classical ground state has all atoms on one sublattice (sublattice A) pointing up (say) and all on sublattice B down.

We can construct a similar state in quantum mechanics, called the Néel state, in which atoms on sublattice A are 'up', i.e. in the $|+S\rangle$ state, and those on sublattice B are 'down', i.e. in the $|-S\rangle$ state. This state however is *not* an eigenstate and so is clearly not the true ground state. Nevertheless we can use it as an approximate ground state, then use SW theory to find the elementary excitations. Finally we can use these to go back and find suitable corrections to the Néel state.

If the total number of deviations from the Néel state is 'small', i.e. if the average value of S^z does not differ greatly from $+S$ or $-S$ for sites on each sublattice respectively, then for sublattice A we can use same transformation as before

$$S^+ \approx (2S)^{\frac{1}{2}}a; \quad S^- \approx (2S)^{\frac{1}{2}}a^+; \quad S^z = S - a^+a$$

(recall that $\langle a^+a \rangle \equiv \langle \hat{n} \rangle$, is 'small').

For sublattice B however, we must use deviations from the $|-S\rangle$ state. The number of deviations is $(S+m)$, e.g. if $S = 6$ and $m = -2$ then the number of deviations from $-S$ is 4, i.e. $S+m$. Clearly in this case more deviations correspond to *higher* S^z. The boson creation operators must therefore lead to states of higher m. We use the notation b^+ and b for the creation and destruction operators on the B sublattice.

Since the operator for the number of deviations is b^+b we must have

$$b^+b = S + S^z$$

so

$$S^z = -S + b^+b \tag{7.9}$$

Likewise a deviation, created by b^+, will now correspond to increasing S^z by one. Hence

$$S^+ \approx (2S)^{\frac{1}{2}}b^+; \tag{7.10}$$

and

$$S^- \approx (2S)^{\frac{1}{2}}b. \tag{7.11}$$

Equations (7.9), (7.10), and (7.11) for the B sublattice correspond to (7.8) for the A sublattice.

Let us use subscript j for the 'up' sublattice A and ℓ for the 'down' sublattice B. ($j+\rho$ is a nearest neighbour of j and so will be on the 'down' sublattice). We can regard the lattice as consisting of N_u 'unit cells', each of which contains 2 atoms, and a sum over one sublattice is equivalent to a sum over the unit cells. Clearly $N_u = N/2$.

The Hamiltonian now becomes

$$\mathcal{H} \approx E_N + \frac{JS}{2}\left(2\sum_j \sum_\rho\right)\left[a_j^+ a_j + b_{j+\rho}^+ b_{j+\rho} + a_j^+ b_{j+\rho}^+ + a_j b_{j+\rho}\right] \tag{7.12}$$

where $E_N = -\dfrac{JNS^2v}{2}$ (J is positive now) is the energy of the Néel state (i.e. the expectation value of $\mathcal{H}$ in the Néel state). The extra factor of 2 is needed since the sum over j is a sum over one sublattice only. Again we have neglected products of four operators.

Now Fourier transform, introducing

$$c_k = \frac{1}{\sqrt{N_u}} \sum_j e^{ikj} a_j \qquad \text{etc.}$$

(similar to the α_k in the ferromagnetic case) and also

$$d_k = \frac{1}{\sqrt{N_u}} \sum_\ell e^{-ik\ell} b_\ell \qquad \text{etc.}$$

The allowed values of k in 1D are now

$$k = \lambda \frac{2\pi}{N_u} \qquad \text{with} \qquad \lambda = 0, 1, \ldots, N_u - 1 = 0, 1, \ldots, \frac{N}{2} - 1.$$

If the lattice spacing in 1D is d, then the actual wavevector q of the original lattice is in units of $1/d$, i.e. $q = k/d$. Because the unit cell of the sublattice is now $2d$ the wavevectors in this spin-wave approximation are in units of $1/(2d)$, i.e. $q = k/(2d)$. This means that for the largest value of k, which is 2π, the largest value of q is π/d. In effect the Brillouin zone is now half the size of the original.

In 2D and 3D the effect is similar but complicated by the fact that the symmetry of the sublattice is in general different to that of the atomic lattice. For example a simple cubic lattice has two f.c.c. sublattices so the Brillouin zone of the sublattice is not related in such a simple way to that of the original lattice.

After Fourier transforming the result is

$$\mathcal{H} = E_N + JS \sum_k \sum_\rho \left[c_k^+ c_k + d_k^+ d_k + e^{ik\rho} c_k^+ d_k^+ + e^{-ik\rho} c_k d_k \right].$$

We define $\gamma_k = \frac{1}{v} \sum_\rho e^{ik\rho}$ ($= \cos k$ in 1D). Therefore

$$\mathcal{H} = E_N + JSv \sum_k \left[c_k^+ c_k + d_k^+ d_k + \gamma_k (c_k^+ d_k^+ + c_k d_k) \right] \qquad (7.13)$$

This form is reminiscent of the $\mathcal{H}$ we obtained in XY model. Note however that these are bosons not fermions. Also clearly k is a 'constant of the motion' here: there is no coupling of different k's or coupling of k and $-k$, unlike in the XY-model.

The eigenvectors will involve linear combinations of c_k and d_k^+. As we noted in the chapter on the XY-model this was first done by Holstein and Primakoff but it is usually known as a Bogoliubov transformation.

Introduce two new Bose operators α_k and β_k:

$$\alpha_k = u_k c_k + v_k d_k^+$$
$$\beta_k = u_k d_k + v_k c_k^+$$

where u_k and v_k are constants which can be taken to be real without loss of generality, so that

$$\alpha_k^+ = u_k c_k^+ + v_k d_k$$
$$\beta_k^+ = u_k d_k^+ + v_k c_k$$

From now on the subscript k will be omitted. Note that both c and d^+ increase the z-component of angular motion by one unit and thus so does α whereas c^+ and d decrease it by one unit and thus so does β.

These operators are required to have the usual Bose commutation relations, namely

$$[\alpha, \alpha^+] = [\beta, \beta^+] = 1$$

with all other pairs of operators from the set $\{\alpha, \alpha^+, \beta, \beta^+\}$ commuting.

We can easily show, using the Bose properties of c and d, that all these commutation relation are satisfied provided that

$$u^2 - v^2 = 1 \tag{7.14}$$

e.g.

$$[\alpha, \alpha^+] = (uc + vd^+)(uc^+ + vd) - (uc^+ + vd)(uc + vd^+)$$
$$= u^2[c, c^+] - v^2[d, d^+] = u^2 - v^2 = 1$$

and

$$[\alpha, \beta] = (uc + vd^+)(ud + vc^+) - (ud + vc^+)(uc + vd^+)$$
$$= u^2(cd-dc) + v^2(d^+c^+-c^+d^+) + uv(d^+d + cc^+ - dd^+ - cc^+)$$
$$= 0 + 0 + uv\left([c, c^+] - [d, d^+]\right) = uv(1 - 1) = 0$$

Now consider the operator which gives the total number of excitations of the two types

$$\alpha^+\alpha + \beta^+\beta = (uc^+ + vd)(uc + vd^+) + (ud^+ + vc)(ud + vc^+)$$
$$= (u^2 + v^2)(c^+c + d^+d) + 2uv(c^+d^+ + cd) + 2v^2.$$

We choose the ratio of coefficients to match the ratio in Eq. (7.13), namely

$$\frac{2uv}{u^2 + v^2} = \gamma \tag{7.15}$$

From Eqs.(7.14) and (7.15) it follows that

$$\frac{u}{v} = \frac{1 \pm s}{\gamma} \tag{7.16}$$

where $s = +\sqrt{1 - \gamma^2}$.   Clearly $s^2 = 1 - \gamma^2$ and $\gamma^2 = 1 - s^2$.

If $\dfrac{u}{v} = \dfrac{1+s}{\gamma}$ and $u^2 - v^2 = 1$ then

$$v^2 \left[\frac{(1+s)^2}{\gamma^2} - 1 \right] = 1$$

$$\frac{v^2}{\gamma^2}[1 + 2s + s^2 - \gamma^2] = 1$$

$$v^2[1 + 2s + s^2 + s^2 - 1] = 1 - s^2$$

$$2v^2 s(1 + s) = (1 - s)(1 + s)$$

$$\therefore \quad 2v^2 = \frac{1 - s}{s}$$

and so

$$u^2 + v^2 = 2v^2 + 1 = \frac{1}{s}. \tag{7.17}$$

Similarly the choice $\dfrac{u}{v} = \dfrac{1-s}{\gamma}$ leads to

$$u^2 + v^2 = -\frac{1}{s}. \tag{7.18}$$

However, as we shall see shortly, this second choice is unphysical and will be discarded.

Hence

$$(c^+c + d^+d) + \gamma_k(c^+d^+ + cd) = \frac{1}{(u^2 + v^2)}(\alpha^+\alpha + \beta^+\beta - 2v^2)$$

and so the Hamiltonian (7.13) can be written

$$\mathcal{H} = E_N + JSv \sum_k \left[\frac{1}{(u^2 + v^2)}(\alpha^+\alpha + \beta^+\beta) - 1 + s \right]. \tag{7.19}$$

Finally we write

$$\mathcal{H} = E_A + \sum_k \varepsilon_k (\alpha^+ \alpha + \beta^+ \beta) \tag{7.20}$$

where the energy of the excitations is

$$\varepsilon_k = JSv\frac{1}{(u^2 + v^2)} = JSvs = JSv\sqrt{1 - \gamma_k^2} \tag{7.21}$$

and

$$E_A = E_N - JSv\sum_k (1 - s) = -\frac{JNS(S+1)v}{2} + JSv\sum_k \sqrt{1 - \gamma_k^2} \tag{7.22}$$

We can now see that choosing the negative sign for $u^2 + v^2$ as in (7.18) rather than the positive sign as in (7.17) would lead to negative excitation energies ϵ_k, and since the excitations are bosons we could obtain an arbitrarily low energy for the system by creating an arbitrary number of them.

The antiferromagnetic spin-waves we have obtained are doubly degenerate with energy $JSv\sqrt{1 - \gamma_k^2}$. The true result is triply degenerate since $S_T = 1$ and there are three degenerate states with $S_T^z = +1, 0, -1$. The operator α_k decreases S_T^z by one unit, while β_k increases it by one unit. There is no operator in spin-wave theory which creates an excitation with no change in S_T^z.

In the 1D case with $S = \frac{1}{2}$ case the exact result for the energy of the excitations is known to be

$$\varepsilon_k = \frac{J\pi}{2}\sin k.$$

while the spin-wave result, using $\gamma_k = \cos k$, $S = \frac{1}{2}$, $v = 2$, has

$$\varepsilon_k = J\sqrt{1 - \cos^2 k} = J\sin k$$

i.e. of correct form but without the $\frac{\pi}{2}$ factor. We also know that in the 1D, $S = \frac{1}{2}$ case the 'spin-wave spectrum' is not a true branch, but rather the lower boundary of a continuum of states.

Nevertheless, even though the 1D, $S = \frac{1}{2}$ case should be the most difficult since it shows the most extreme quantum effects, the results are rather good. For higher S and higher dimension the spin-wave results are even more satisfactory.

We can use our results (7.20), (7.21), and (7.22) to obtain an estimate of how different the energy of the true ground state is from that of the Néel state. The energy of the ground state is the value of (7.20) with $\alpha_k^+\alpha_k$ and $\beta_k^+\beta_k$ put to zero, since these operators count the number of *excited* bosons (spin-waves). This gives

$$E_A \;=\; E_N - \frac{NJv}{2}S + JSv \sum_k \sqrt{1 - \gamma_k^2}.$$

The sum over k can be converted to an integral which has the value $\frac{2}{\pi} \times \frac{1}{2}$ in 1D and can be evaluated numerically in higher D. (The factor of $\frac{1}{2}$ comes from the fact that the summation over k runs over $\frac{N}{2}$ terms.) In 1D, for $S = \frac{1}{2}$, the result is

$$\frac{E_A}{N} \;=\; -0.75J + \frac{J}{\pi} \quad (\text{or } -0.431690J \text{ to 6 decimal places})$$

while the exact result is known to be $-0.443147J$ (to 6 decimal places) which is in quite good agreement. (Note that the energy of the Néel state itself is $\frac{E_N}{N} = -0.25$, which is much less accurate!).

One can also calculate $\langle S_j^z \rangle$ in the ground state in the SW approximation since $S_j^z = S - a_j^+ a_j$ (for the up sublattice). This is done using the inverse Fourier transform $\sum_j \langle a_j^+ a_j \rangle = \sum_k \langle c_k^+ c_k \rangle$, the inverse of the Bogoliubov transformation $c_k = u\alpha_k - v\beta_k^+$ and the fact that $\langle \alpha_k^+ \alpha_k \rangle = \langle \beta_k^+ \beta_k \rangle = 0$ in the SW ground state.

Putting $\langle \delta S_i^z \rangle = S - \langle S_i^z \rangle$, then for $S = \frac{1}{2}$ the results are

$$\langle \delta S_i \rangle$$

	$\langle \delta S_i \rangle$
3D (simple cubic)	0.078
2D (square)	0.20
1D	∞

The result for 1D is clearly unphysical, and is associated with the fact that there is no long-range order. However for 2D and 3D the results are in quite good agreement with much more sophisticated calculations.

References

1. Anderson, P.W.: Phys. Rev. **86**, 694 (1952)
2. Oguchi, T.: Phys. Rev. **117**, 117 (1960)

Chapter 8
Numerical Finite-Size Calculations

Abstract Numerical methods have been used extensively for studies of quantum spin systems. In subsequent chapters we shall describe some of these methods, which make use of sophisticated approximation techniques developed in many areas of quantum many-body theory. However, in this chapter we first look a very simple numerical method that does not involve any theory other than the assumption that the properties of small systems change smoothly as the size of the system increases to the limit of infinite numbers of atoms. The technique is to take a small finite-sized system with N atoms, where N can be any number from 2 upwards. For these small systems, every basis state can be written down explicitly and all the matrix elements of the Hamiltonian in the basis can be calculated. The matrix is then diagonalised numerically and hence the eigenstates and eigenvalues calculated. Usually the largest N is of the order of 20 for full diagonalisation and 40 for partial diagonalisation. Clearly the ground state energy and the energies of the elementary excitations are obtained directly. Also the partition function can be constructed directly for full diagonalisation because all the eigenvalues are known. This opens the way to study of the non-zero temperature properties of these systems.

8.1 Introduction

In previous chapters we considered two exact 1D methods:

1. The Bethe Ansatz applied to the $S = \frac{1}{2}$ Heisenberg chain.
2. The method of Jordan and Wigner applied to the $S = \frac{1}{2}$ XY chain.

These exactly solvable systems are known as *integrable* systems. They comprise only a few very specialised systems, which are nevertheless extremely important because much more detailed information is available for them than for other systems.

We have also considered one approximate method, namely spin-wave theory, especially for antiferromagnets which is valid in all dimensions. Spin-wave theory is useful in practice but is not easy to improve in a systematic way.

In this section we describe another approximate method which applies very generally and which is systematic. This method is useful for estimating the ground

Parkinson, J.B., Farnell, D.J.J.: *Numerical Finite-Size Calculations*. Lect. Notes Phys. **816**, 89–97 (2010)
DOI 10.1007/978-3-642-13290-2_8

state properties, i.e. for systems with $T = 0$. It also works well for medium and high T, but is least accurate for low $T (\neq 0)$. Again we shall mainly consider antiferromagnets.

We shall describe the essential features of the method. However, because it has been very widely used it is only possible here to give one or two examples of its application which we hope will give an indication of its power.

Suppose we wish to find the ground state energy for a system of N atoms where $N \to \infty$. As noted above, most systems are not integrable, examples being

a. Systems with $S \geq 1$.
b. Systems with interactions which are not nearest-neighbour.
c. Systems with other types of interaction between spins such as biquadratic $(\mathbf{S}_i \cdot \mathbf{S}_j)^2$.
d. Systems with dimension higher than 1.

For many of these cases a powerful method is to calculate results numerically for small systems, typically $N \lesssim 40$, for several different values of N and then try to extrapolate $N \to \infty$.

The pioneering work was done by Bonner and Fisher (1964) [1] who considered the 1D (chains and rings) XXZ model with $S = \frac{1}{2}$ and a magnetic field.

$$\mathcal{H} = -2J \sum_j \Delta \left[S_j^z S_{j+1}^z + (S_j^x S_{j+1}^x + S_j^y S_{j+1}^y) \right] - B \sum_j S_j^z. \qquad (8.1)$$

(Note the factor of 2 compared to earlier chapters.) This is a system which can be dealt with by Bethe Ansatz (BA) and so the accuracy of some of the numerical results, such as the ground state energy, can be checked directly with the exact result. Other properties of the ground state, such as correlations between different spins, which cannot be easily obtained by the BA can now be estimated numerically. Also results for non-zero T can be found. Mostly it has been used for 1D systems as it is more difficult to treat 2D systems that are large enough to make extrapolations meaningful. However, for some systems accurate results in 2D for the ground state have been found and some of these are discussed in the final chapter. For the same reason there are virtually no results in 3D.

8.2 A Simple Example

The method involves direct diagonalisation of the full Hamiltonian for a short chain (open ends) or ring (periodic boundary conditions). We illustrate the method for the Heisenberg Hamiltonian, i.e. (8.1) with $\Delta = 1$ and $B = 0$, for a ring of $N = 4$, $S = \frac{1}{2}$ atoms. We also put $J = 1$ for simplicity. Hence

$$\mathcal{H} = 2 \sum_j S_j \cdot S_{j+1} = 2 \sum_j \left[S_j^z S_{j+1}^z + \frac{1}{2}(S_j^- S_{j+1}^+ + S_j^+ S_{j}^-) \right] \qquad (8.2)$$

Here $S_T^z = \sum_j S_j^z$ is a good quantum number. A complete set of basis states has $2^N = 16$ states, which in the usual notation where plus denotes $+\frac{1}{2}$, and minus denotes $-\frac{1}{2}$ are

$$|++++> \qquad S_T^z = 2$$

$$\left.\begin{array}{l} |+++-> \\ |++-+> \\ |+-++> \\ |-+++> \end{array}\right\} \quad S_T^z = 1$$

$$\left.\begin{array}{l} |++--> \\ |+--+> \\ |--++> \\ |-++-> \\ |+-+-> \\ |-+-+> \end{array}\right\} \quad S_T^z = 0$$

The remaining basis states for $S_T^z = -1$ and $S_T^z = -2$, follow the same pattern.

We now calculate the effect of operating with $\mathcal{H}$ on each of the basis states, e.g.

$$\mathcal{H}|++-+> = \tfrac{1}{2}|+-++> +\tfrac{1}{2}|+++->,$$
$$\mathcal{H}|++--> = \tfrac{1}{2}|+-+-> +\tfrac{1}{2}|-+-+>,$$
$$\mathcal{H}|+-+-> = -|+-+-> +\tfrac{1}{2}|-++->$$
$$\qquad \tfrac{1}{2}|++--> +\tfrac{1}{2}|+--+>$$
$$\qquad \tfrac{1}{2}|--++>,$$

and eventually we obtain the following representation of $\mathcal{H}$

$$\mathcal{H} = \begin{pmatrix} 1 & & & \\ & \begin{smallmatrix} 0 & \frac{1}{2} & 0 & \frac{1}{2} \\ \frac{1}{2} & 0 & \frac{1}{2} & 0 \\ 0 & \frac{1}{2} & 0 & \frac{1}{2} \\ \frac{1}{2} & 0 & \frac{1}{2} & 0 \end{smallmatrix} & & \\ & & \begin{smallmatrix} 0 & 0 & 0 & 0 & \frac{1}{2} & \frac{1}{2} \\ 0 & 0 & 0 & 0 & \frac{1}{2} & \frac{1}{2} \\ 0 & 0 & 0 & 0 & \frac{1}{2} & \frac{1}{2} \\ 0 & 0 & 0 & 0 & \frac{1}{2} & \frac{1}{2} \\ \frac{1}{2} & \frac{1}{2} & \frac{1}{2} & \frac{1}{2} & -1 & 0 \\ \frac{1}{2} & \frac{1}{2} & \frac{1}{2} & \frac{1}{2} & 0 & -1 \end{smallmatrix} & \\ & & & \begin{smallmatrix} 0 & \frac{1}{2} & 0 & \frac{1}{2} \\ \frac{1}{2} & 0 & \frac{1}{2} & 0 \\ 0 & \frac{1}{2} & 0 & \frac{1}{2} \\ \frac{1}{2} & 0 & \frac{1}{2} & 0 \end{smallmatrix} \\ & & & & 1 \end{pmatrix}$$

where blank regions contain only 0s. Note the block-diagonalisation due to S_T^z being a good quantum number.

Clearly this matrix can now be easily diagonalised numerically to find all the eigenvalues, the lowest of which is the ground state energy.

The system also has translational symmetry because of the periodic boundary conditions. This means that we can Fourier transform in terms of wavevectors k. For this small system these can be written directly as a new basis, which for the $S_T^z = 0$ subspace has the form

$$\phi_1 = \tfrac{1}{\sqrt{2}}(|+-+-> + |-+-+>) \qquad\qquad\qquad k = 0$$

$$\phi_2 = \tfrac{1}{\sqrt{2}}(|+-+-> - |-+-+>) \qquad\qquad\qquad k = \pi$$

$$\phi_3 = \tfrac{1}{2}(|++--> + |-++-> + |--++> + |+--+>) \quad k = 0$$

$$\phi_4 = \tfrac{1}{2}(|++--> + i|-++-> - |--++> - i|+--+>) \; k = \tfrac{\pi}{2}$$

$$\phi_5 = \tfrac{1}{2}(|++--> - |-++-> + |--++> - |+--+>) \quad k = \pi$$

$$\phi_6 = \tfrac{1}{2}(|++--> - i|-++-> - |--++> + i|+--+>) \; k = \tfrac{3\pi}{2}$$

with similar states for the $S_T^z = \pm 1$ subspace. (The $S_T^z = \pm 2$ subspaces only have one state in so this is automatically a $k = 0$ state.)

Note that these states are of the form

$$\phi = \frac{1}{\sqrt{N_T}} \sum_j e^{ikj} (T)^j \psi$$

where ψ is one of original basis states, T translates by 1 unit and N_T is the number of distinct states which are related to ψ by translation. This greatly reduces the size of the matrix to be diagonalised. One can also make use of the reflection symmetry to create a basis whose states have only real coefficients, e.g. $(\phi_4 + \phi_6)$ and $i(\phi_4 - \phi_6)$ instead of ϕ_4 and ϕ_6, which enables the Hamiltonian matrix have purely real entries and thus be easier to handle numerically. The ground state wave function is given by "$\phi = (\sqrt{2}\phi_1 - \phi_3)/\sqrt{3}$, which gives an overall ground-state energy eigenvalue of -4."

8.3 Results in 1D

In Table 8.1 and Fig. 8.1 we show the results that Bonner and Fisher obtained for the ground state energy per atom of the Heisenberg antiferromagnetic chain using this method. (Note the factor of 2 in Eq. (8.1).) As can be seen the results extrapolate very accurately as straight lines when plotted as a function of $\frac{1}{N^2}$ rather than $\frac{1}{N}$. Also note the separate lines for chains with odd and even numbers of atoms, both of which converge to the exact value of $\frac{1}{2} - 2\ln 2$ as $N \to \infty$.

Clearly the results in this case where the exact result is known are very good, and for other 1D systems which are not exactly soluble the results are generally excellent.

Table 8.1 Finite-size results for the ground-state energy of spin-half Heisenberg chains of (odd/even) length N with periodic boundary conditions (note the factor of 2 compared to earlier results of Bethe Ansatz)

N	Even	N	Odd
4	−1.00000	3	−0.50000
6	−0.93425	5	−0.74721
8	−0.91277	7	−0.81577
10	−0.90309	9	−0.84384
12	−0.8979	11	−0.85799
∞	−0.88629	∞	−0.88629

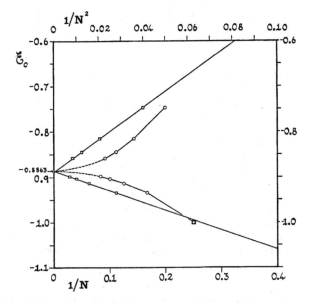

Fig. 8.1 Bonner and Fisher's [1] original figure for the ground state energy of the linear Heisenberg antiferromagnetic chain. (Antiferromagnetic ground-state energies versus $1/N$ (*circles*) and versus $(1/N)^2$ (*squares*) for pure Heisenberg rings ($\gamma = 1$).) (Reprinted with permission from Bonner and Fisher [1]. Copyright 1964 by the American Physical Society)

Because all the eigenvalues can be calculated for small systems of size N it is quite straightforward to calculate the partition function:

$$Z_N = \sum_i e^{-\beta E_i},$$

and from this we can calculate any other thermodynamic quantity such as specific heat. Bonner and Fisher's result is shown in Fig. 8.2. As can be seen the convergence is good for $T \gtrsim 1$.

At the time of Bonner and Fisher's paper there were very few $T \neq 0$ results available, even for $S = \frac{1}{2}$. Later work by Yang and Yang (1966) [2–4] and Takahashi (1971) [5] enabled some of these quantities to be calculated exactly using the Bethe Ansatz for the $S = \frac{1}{2}$ chains. The numerical method is much more general, however,

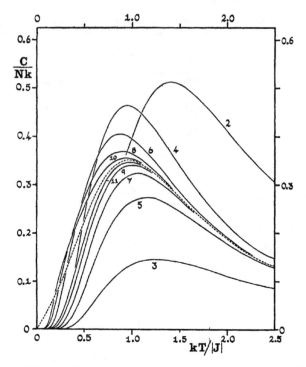

Fig. 8.2 Bonner and Fisher's [1] original figure for the specific heat of the linear Heisenberg anti-ferromagnetic chain. (Variation of specific heat with temperature for antiferromagnetic Heisenberg chains: finite N, *solid lines*; estimated limit $N = \infty$, *dashed line*.) (Reprinted with permission from Bonner and Fisher [1]. Copyright 1964 by the American Physical Society)

and works well for many of the systems mentioned at the beginning of this chapter, e.g. $S \geq 1$, for which the Bethe Ansatz method is not applicable.

Another important result which was confirmed by this method is the existence of an energy gap between the ground state and the first excited state of linear chains with an isotropic Heisenberg interaction. This gap is known to be zero, i.e. there is no gap, in the limit $N \rightarrow \infty$ for $S = \frac{1}{2}$. However, in 1982 Haldane [6, 7] predicted that for $S = 1$ (and any other integer value of S) the gap would be non-zero. Haldane's method involved a transformation of the chain of separate atoms into a continuum limit and then using field theoretical methods. He was not able to prove the result exactly and initially people were surprised and somewhat sceptical.

However numerical results by Botet and Jullien (1983) [8], using the above techniques rapidly produced evidence that the result was correct. Figure 8.3 is from Parkinson and Bonner (1985) [9] and gives results for longer chains. If the gap for $S = 1$ were zero then the points would have to lie on a curve similar to the one shown with dots and dashes, which seems unlikely.

Much later, in 1993, the value of the gap in energy was obtained to very great accuracy by the density matrix renormalisation group method (DMRG) by White

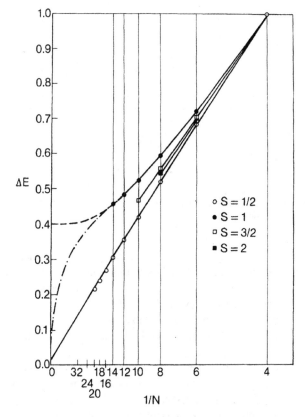

Fig. 8.3 Plot of the energy gap of the linear Heisenberg antiferromagnetic chain for different S

and Huse [10] and the value is 0.4105(1) in units of the exchange J. The DMRG is another quite different numerical technique which works extremely well in 1D but is not easy to extend to 2D.

This non-zero gap for integer spin S is evidence of a completely different type of ground state which is disordered and has correlations which decay exponentially. The systems with $S =$ integer $+\frac{1}{2}$ including the Bethe Ansatz soluble case $S = \frac{1}{2}$ are believed to have ordered ground states in which correlations decay algebraically.

8.4 Results in 2D

As mentioned earlier, the method can be used in 2D. For a square lattice one can choose small square sections containing N atoms and with periodic boundary conditions. If the square section has sides parallel to the axes of the square then the number of suitable square sections is five, namely 2×2, 3×3, 4×4, 5×5 and 6×6. 6×6 would have $N = 36$, and which is almost at the limit of what can be

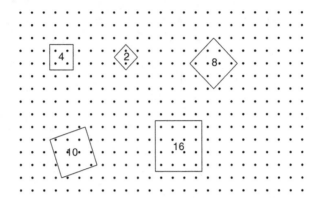

Fig. 8.4 *Square sections* of an infinite square lattice with $N = 2, 4, 8, 10, 16$ which would tile the entire lattice if repeated indefinitely

handled numerically for the ground state, e.g., via the Lanczos method. However, an ingenious way around this, introduced by Oitmaa and Betts (1977) [11], is to use square sections with periodic boundary conditions where the edges of the square are not parallel to the axes. This is shown in Fig. 8.4 for sections with $N = 2, 4, 8, 10$ and 16. Similar constructions are also available for $N = 18, 20$, etc. Using these one can plot the ground state energy and other quantities as a function of $1/N$ or $1/N^2$ just as for 1D systems.

The limiting factor in all of these calculations is the very rapid growth in the size of the basis as N increases. For $S = \frac{1}{2}$ systems the number of basis states is 2^N so for $N = 20$, $2^{20} = 1048576$. For $S = 1$ number of states is 3^N etc. The matrices are somewhat smaller than this because of translational symmetry and reflection symmetry in 1D and also rotational symmetry in 2D. Nevertheless, the largest values of N that can be fully diagonalised in 1D are approximately of order 20 for $S = \frac{1}{2}$, 14 for $S = 1$ and 10 for $S = \frac{3}{2}$. In 2D the maximum N are slightly greater because there is more symmetry but, of course, it is $\sqrt{N}$ which determines the size of the section in 2D and this is much smaller. Larger systems may be considered for partial diagonalisation, e.g., for the ground state and low-lying excited states. Indeed, results of numerical finite-size calculations for the ground states of 2D antiferromagnets (referred to also as "exact diagnalisations") are presented in the final chapter of this book. It is clear that in 3D lattices one cannot use large enough sections for the method to be useful at the moment.

References

1. Bonner, J.C., Fisher, M.E.: Phys. Rev. **135**, A640–A658 (1964)
2. Yang, C.N., Yang, C.P.: Phys. Rev. **150**, 321–327 (1966)
3. Yang, C.N., Yang, C.P.: Phys. Rev. **150**, 327–339 (1966)
4. Yang, C.N., Yang, C.P.: Phys. Rev. **151**, 258–264 (1966)
5. Takahashi, M.: Prog. Theor. Phys. **46**, 401–415 (1971)

6. Haldane, F.D.M.: Phys. Lett. **93A**, 464–468 (1983)
7. Haldane, F.D.M.: Phys. Rev. Lett. **50**, 1153–1156 (1983)
8. Botet, R., Jullien, R.: Phys. Rev. B **27**, 613–615 (1983)
9. Parkinson, J.B., Bonner, J.C.: Phys. Rev. B **32**, 4703–4724 (1985)
10. White, S.R., Huse, D.A.: Phys. Rev. B **48**, 3844 (1993)
11. Oitmaa, J., Betts, D.D.: Can. J. Phys. **56**, 897–901 (1978)

Chapter 9
Other Approximate Methods

Abstract Exact diagonalisations of the type presented in the previous chapter are typically limited to relatively small numbers of lattice sites. In this chapter approximate methods that may be applied to larger systems are discussed. The variational method is discussed as one method of simulating the properties of quantum spin systems, although an exact enumeration of the basis states in a given sector becomes prohibitive in terms of computational cost for larger lattices. Monte Carlo simulation allows us to treat larger lattices using the variational method. This discussion leads on to a more general description of the quantum Monte Carlo simulation of quantum spin systems. Finally, the topics of perturbation theory and of series expansions are explored. Series expansions obey the linked cluster theorem and so yield results valid in the infinite-lattice limit from the outset. The spin-half Heisenberg model on the linear chain is used as a test-case for all of these methods. We show that even simple applications of these methods give improved results for the ground-state energy compared to the classical result.

9.1 Introduction

The finite-size calculations, described in the previous chapter, are only one way of obtaining results numerically. The coupled-cluster method, described in detail in the next chapter is also a numerical method which deals directly with the infinite lattice. In this chapter we mention the variational method and its stochastic version, the Variational Monte Carlo method, the Green Function Monte Carlo method and also perturbation theory/series expansions. Some of these, e.g., series expansions, deal directly with the infinite lattice, although others such as Monte Carlo are generally applied to a finite lattice and the results then extrapolated in the infinite-lattice limit.

9.2 Variational Method

The variational method, widely used in theoretical physics, can also be applied to 2D and 3D quantum spin systems in a simple and straightforward manner. An important point to note about the variational method is that the approximate bra

Parkinson, J.B., Farnell, D.J.J.: *Other Approximate Methods*. Lect. Notes Phys. **816**, 99–108 (2010)
DOI 10.1007/978-3-642-13290-2_9   © Springer-Verlag Berlin Heidelberg 2010

and ket states are Hermitian conjugates of each other (unlike in the coupled cluster method described later) so that the calculated ground-state energy is a strict upper bound on the true value. If the result depends upon some parameter in the wave function then minimising with respect to this parameter, i.e. varying it to find the minimum, can lead to a good approximation to the true value. The ground state wave function is less reliably obtained as some important features may have been neglected in the initial trial wave function.

For example, in an antiferromagnetic spin system we usually know that the ground state is a state with total spin equal to 0. The true ground state must consist of a linear combination of states in the set $\{I\}$ which is the set of all Ising states for which $S_T^z \equiv \sum_i S_i^z = 0$. Thus the ground state wave function is

$$|\Psi\rangle = \sum_I c_I |I\rangle. \tag{9.1}$$

The ground-state energy is given by the Schrödinger equation as normal

$$H|\Psi\rangle = E_g|\Psi\rangle$$

and applying $\langle\Psi|$ on the left gives

$$E_g = \frac{\langle\Psi|H|\Psi\rangle}{\langle\Psi|\Psi\rangle}$$

or

$$E_g = \frac{\sum_{I_1,I_2} c_{I_1}^* c_{I_2} \langle I_1|H|I_2\rangle}{\sum_I |c_I|^2}. \tag{9.2}$$

Now we make an approximation to the coefficients in such a way that there is one or more parameters in the coefficients which can be varied.

An example is the choice for a spin-half system given by

$$c_I = u_{ij}(p_i^\uparrow p_j^\downarrow + p_i^\downarrow p_j^\uparrow), \tag{9.3}$$

where the indices i and j run over all lattice sites. $p^\uparrow$ and $p^\downarrow$ are projection operators for the 'up' and 'down' states of the spins, where

$$p^\uparrow|\uparrow\rangle = 1|\uparrow\rangle \quad p^\uparrow|\downarrow\rangle = 0|\downarrow\rangle$$
$$p^\downarrow|\uparrow\rangle = 0|\uparrow\rangle \quad p^\downarrow|\downarrow\rangle = 1|\downarrow\rangle.$$

In terms of the normal spin operators

$$p^\uparrow \equiv \frac{1}{2} + S^z : \qquad p^\downarrow \equiv \frac{1}{2} - S^z,$$

and the $\{u_{ij}\}$ are parameters.

Note that it is slightly easier to deal with the equivalent spin system after a unitary rotation of the local axes of the spin by 180° on one sublattice. This changes the Hamiltonian slightly, but crucially means that the c_I coefficients must be positive from the Marshall-Peierls sign rule. An example is the choice for a spin-half system (after rotation of the local spin axes) given by

$$c_I = u_{ij}(p_i^\uparrow p_j^\uparrow + p_i^\downarrow p_j^\downarrow), \qquad (9.4)$$

where the indices i and j run over all lattice sites.

A possible choice of these parameters is the 'two-site' approximation known as the Jastrow Ansatz, in which the u_{ij} are all zero if i and j are not nearest neighbours. A simple version of this is to set u_{ij} equal to some number α when i and j are nearest-neighbours and to zero in all other cases. There is now a single parameter, α, to be adjusted in the final result.

For small enough lattices, we enumerate all of the states in the Ising basis either by hand (or for larger lattices computationally), and so we are able to calculate all of the contributions to the ground-state energy of Eq. (9.2). We obtain a value for the ground-state energy in terms of α and then minimise with respect to it. Hence, we obtain a 'variational' value for the ground-state energy. Because the ground-state energy thus found is an upper bound of the 'true' value, this is very useful both as an estimate of the ground-state energy in itself and also as a test for other approximate methods. If another approximate method produces a ground-state energy higher than the simple variational estimate then it is unsatisfactory.

Finally, we may find estimates of other ground-state expectation values easily as we have direct access to the (approximate) wave function written in terms of expansion coefficients with respect to the Ising basis.

9.3 Variational Monte Carlo Method

However, we may not be able to enumerate all Ising states in the basis even using very intensive computational approaches (and in the relevant ground-state subspace) for very large lattices, i.e., those of very many spins. This is particularly true for 2D and 3D lattices and for higher spin quantum number. In this case we use a method called Monte Carlo simulation in order to carry out the enumeration of states approximately. Clearly, not all states are equal in their contribution to the ground-state energy and this method allows us to choose those states that are 'most important.' Furthermore, the error of the estimate of ground-state expectation values decreases in a statistically well-understood manner with the length of the simulation.

We begin the Monte Carlo treatment of the variational problem by rewriting the
ground-state energy of Eq. (9.2) in the following form:

$$E_g = \frac{\sum_{I_1,I_2} c_{I_1}^* c_{I_2} \langle I_1|H|I_2 \rangle}{\sum_I |c_I|^2}$$

$$= \frac{\sum_{I_1} |c_{I_1}|^2 \sum_{I_2} \frac{c_{I_2}}{c_{I_1}} \langle I_1|H|I_2 \rangle}{\sum_I |c_I|^2}$$

$$\Rightarrow \quad E_g = \sum_I P(I) E_L(I). \tag{9.5}$$

$P(I)$ is interpreted as a probability for the state I and is given by

$$P(I) = |c_I|^2 / \sum_J |c_J|^2.$$

(The index J represents a sum over all Ising basis states, though it will disappear as
we shall see shortly). The local energy with respect to the state I is given by

$$E_L(I) = \sum_{I_2} \frac{c_{I_2}}{c_I} \langle I|H|I_2 \rangle.$$

We start from a given state in the relevant ground-state Ising basis and we define an
acceptance probability $A(I \rightarrow I')$ from state I to state I' given by

$$A(I \rightarrow I') = \min \left[1, \frac{P(I')K(I)}{P(I)K(I')} \right], \tag{9.6}$$

where $K(I)$ indicates the number of states accessible from state I via the 'off-
diagonal' terms in the Hamiltonian. (The denominator $\sum_J |c_J|^2$ in both $P(I)$ and
and $P(I')$ is identical and so cancels in Eq. (9.6).) We choose a state I' to be one
of these $K(I')$ states and now use the Metropolis algorithm in order to decide if we
should accept this new state I' or if we should stay with the old one, I. We generate
a random number η from a uniform distribution in the range 0 to 1 and we test if our
acceptance probability $A(I \rightarrow I')$ is greater than η. If it is then we accept the new
state and otherwise we retain the original state. It is this element of randomness or
chance that leads to the name of the method (i.e., Monte Carlo) for obvious reasons.

We now repeat the process for the new state I' in order to form yet another state
I'', and so on. As is common in Monte Carlo simulations we repeat this process
many times and we form an average of the local energies (and 'local' estimates of
all other ground-state expectation values similarly) as we go along. Again, we note

that the ground-state energy again forms an upper bound on the 'true' ground-state energy, but within the limits of the error bars due to the Monte Carlo simulation in this case. The estimate of the ground-state energy may be minimized computationally for the variational Monte Carlo estimate presented here. This is simple to achieve in practice and a robust estimate of the ground-state energy may be formed in a straightforward manner for the nearest-neighbour two-spin approximation. This process becomes more difficult as we include longer-range correlations and higher-order terms in the approximation for the c_I coefficients. However, we would expect the accuracy of the results to increase also as we add in more such terms and, again, we would obtain an upper bound on the ground-state energy.

It is interesting to apply the variational Monte Carlo (VMC) method to the spin-half one-dimensional antiferromagnetic ($J>0$) Heisenberg model on a chain of finite length, which can be solved by the exact diagonalisation method as described above. For example, we obtain an estimate of the ground state energy of $E_G/N = -0.423729(3)J$ ($\alpha = 1.773$) for the Heisenberg model on a chain of $N = 20$ sites with periodic boundary conditions. Note that α is the strength of the nearest-neighbour correlations in u_{ij} in Eq. (9.4) and that $M = 10^{10}$ Monte Carlo iterations were used in this simulation. This result lies above and is reasonably close to the exact diagonalisation result of $E_G/N = -0.445219J$ for the $N = 20$ chain with periodic boundary conditions, which is itself not too far from the exact Bethe ansatz result of $E_G/N = 1/4 - \ln 2 \ (= -0.443147)$ in the infinite-lattice limit. Indeed, we see that 89% of the correlation energy of the $N = 20$ Heisenberg chain has been captured by using this simple two-body nearest-neighbour Ansatz. Furthermore, this result also illustrates the fact mentioned earlier that variational methods provide an upper bound on the ground-state energy of a system because the ket and bra states are always Hermitian conjugates of each other. We may also apply VMC to much larger lattices than exact diagonalisations because the computational cost needed to carry out VMC is much lower. However, caution should be used when interpreting the results of variational studies because they might actually provide quite a poor approximation to the 'true' ground-state wave function while still yielding deceptively good results for the ground-state energy.

Despite this, however, it is true to say that variational methods have an important role to play in understanding quantum spin systems, especially in providing upper bounds on the ground-state energy which can then be compared with results of other approximate methods. In the next chapter we shall describe a method called the coupled cluster method (CCM) that allows us to systematically refine the approximation level to include higher order terms in the wave function. The CCM is strictly speaking a bi-variational method, and since the bra and ket states are not explicitly constrained to be Hermitian conjugates of each other, it does not yield an upper bound to the energy. However, this negative aspect is offset by many positive aspects, not-the-least that one may obtain results for the infinite lattice ($N \rightarrow \infty$) from the outset.

9.4 The Green Function Monte Carlo Method

There are many forms of Monte Carlo in science [1–6] and its use is ubiquitous. Thus, even in the fairly restricted area of quantum spin systems, the variational Monte Carlo method is only one of a number of approaches that use Monte Carlo techniques. Another example, used for calculating zero temperature properties, is the Green function Monte Carlo method. This provides an estimate of the true ground-state energy and other properties directly and so there is no need to minimise the ground-state energy with respect to any parameters of the wave function.

It uses 'power iteration' of the Hamiltonian in which one repeatedly operates with the Hamiltonian on an initial starting state $|I_0\rangle$ (in the relevant ground-state subspace). For a large number M of operations $H^M |I_0\rangle \overset{M \to \infty}{\longrightarrow} K |\Psi\rangle$, where K is a constant and $|\Psi\rangle$ is the eigenstate of H which has the eigenvalue with the largest magnitude, normally the ground-state. This method is frequently used in exact diagonalizations of finite-sized systems in order to isolate the ground-state wave function and its energy. Often the Lanczos technique can also be used to drastically reduce the number of iterations compared to the direct 'power" iteration method.

However, for the Green function Monte Carlo, this property of power or direct iteration turns out to be quite useful. This is because each time we apply the Hamiltonian new Ising states in the approximate wave function are created. The Green function method represents these states by a set of 'walkers' where each walker has transition probabilities $A(I \to I')$ to go from state I to state I' and associated local energies. The estimate of the ground-state energy is given by the average of the local energies over all walkers for a given number of iterations M. Again, each move is accepted or rejected randomly and so we obtain a 'random walk' for each of the individual walkers through the set of basis states.

For Green function Monte Carlo, however, we also need to know the signs of the expansion coefficients c_I beforehand in order to ensure that the transition probabilities are always positive; if we utilize signs that are wrong then we will sample the underlying probability distributions incorrectly and so our estimates will also be incorrect. Fortunately, there exist a number of such rules for the signs of the expansion coefficients, the most famous of which is the Marshall-Peierls sign rule [7] for the Heisenberg model for bipartite lattices. Bipartite lattices are those lattices such as that linear chain, square and cubic latices that can be decomposed into two neighbouring sublattices. The estimates of the ground-state properties of lattice quantum spin systems thus obtained using Green function Monte Carlo present the most important of the approximate methods for 2D (and to lesser extent 3D) unfrustrated systems. However, the application of Monte Carlo is severely limited by the presence of the 'sign problem' for frustrated spin systems. Frustration is an effect in which different terms in the Hamiltonian compete. Classically, this means that the ground-state energy per bond is lower than that of a comparative unfrustrated system. By contrast, perhaps the best evidence of frustration occurs in the analogous quantum system is that no such sign rule can be created. Examples of frustrated systems are the triangular lattice antiferromagnet and the J_1–J_2 model with antiferromagnetic nearest- and next-nearest-neighbour terms of bond strengths J_1 and J_2, respectively. Some of these systems will be considered in more depth later on. However, we note that variational Monte Carlo may still be applied for those cases in which no sign rule exists. Furthermore, a more sophisticated Monte Carlo approach called fixed-node Monte Carlo may be employed in these cases also. However, a description of this method lies beyond the scope of this text.

9.5 Perturbation Theory

There are many other approximate methods which have been applied to quantum spin systems. We finish by briefly mentioning one of these, the perturbation method.

The basic idea here is to divide $\mathcal{H}$ into two parts and treat one part as a perturbation, even if it is not very small. For example, a system with anisotropic exchange, called the 'XXZ-Model', in any number of dimensions has Hamiltonian

$$\mathcal{H} = \frac{J}{2} \sum_i \sum_\rho \left[S_i^z S_{i+\rho}^z + \gamma \left(S_i^x S_{i+\rho}^x + S_i^y S_{i+\rho}^y \right) \right]$$

where ρ runs over all nearest neighbours on the opposite sublattice and γ is the anisotropy parameter. Writing this as

$$\mathcal{H} = \mathcal{H}_0 + \mathcal{H}_1$$

where

$$\mathcal{H}_0 = \frac{J}{2} \sum_i \sum_\rho S_i^z S_{i+\rho}^z$$

and

$$\mathcal{H}_1 = \frac{J}{2} \gamma \sum_i \sum_\rho (S_i^x S_{i+\rho}^x + S_i^y S_{i+\rho}^y)$$

$\mathcal{H}_0$ is a simple Ising model and its exact ground state is the antiferromagnetic Néel state. Clearly for small values of γ, $\mathcal{H}_1$ is small and can be treated as a perturbation. Hence, using perturbation theory, one can calculate corrections to the ground state energy. One can also calculate correlations. The method has the advantage that it works for all S and all dimensions including 3D.

As an illustration, we calculate the first few orders of the (Rayleigh-Schrödinger) perturbation series for the energy $E_G = E_0 + \gamma E_1 + \gamma^2 E_2 + \cdots$ for the $S = \frac{1}{2}$ case explicitly. It is useful to rotate the the axes by $180°$ on one sublattice so that notionally the classical Néel (unperturbed) state $|0\rangle$ has all spins pointing upwards, and this has the effect that $H_0 = -\frac{J}{2} \sum_i \sum_\rho S_i^z S_{i+\rho}^z$ and $H_1 = -\frac{J}{4}\gamma \sum_i \sum_\rho (S_i^+ S_{i+\rho}^+ + S_i^- S_{i+\rho}^-)$. The zeroth-order term in the series for the energy is now simply:

$$E_0 = \langle 0|H_0|0 \rangle = \frac{-NnJ}{8},$$

where n is the number of nearest neighbours (1D: $n = 2$; 2D: $n = 4$; and, 3D: $n = 6$).

The first-order correction is given by $E_1 = \langle 0|H_1|0 \rangle$. H_1 acting on $|0\rangle$ produces pairs of nearest-neighbour down spins for each lattice site i. Denoting one such state as $|i, i + \rho\rangle$ and noting that

$$\langle 0|i, i + \rho\rangle = 0$$

we see immediately that

$$E_1 = 0. \tag{9.7}$$

The second-order term in the series for the energy is given by

$$E_2 = \frac{1}{2} \sum_i \sum_\rho \frac{|\langle i, i + \rho|H_1|0 \rangle|^2}{(E_0 - E_0^{(i,i+\rho)})}, \tag{9.8}$$

where the factor of $\frac{1}{2}$ is to avoid over-counting, and where

$$E_0^{(i,i+\rho)} = \langle i, i + \rho|H_0|i, i + \rho\rangle = (E_0 + J) \quad \text{in 1D}$$
$$= (E_0 + 3J) \quad \text{in 2D}$$
$$= (E_0 + 5J) \quad \text{in 3D}.$$

Furthermore, $\langle i, i + \rho|H_1|0 \rangle = J\gamma/2$, so

$$E_2 = \frac{Nn}{2} \times \frac{(J/2)^2}{(E_0 - E_0 - J)} = -\frac{NJ}{4} \quad \text{in 1D}$$
$$= \frac{Nn}{2} \times \frac{(J/2)^2}{(E_0 - E_0 - 3J)} = -\frac{NJ}{6} \quad \text{in 2D}$$
$$= \frac{Nn}{2} \times \frac{(J/2)^2}{(E_0 - E_0 - 5J)} = -\frac{3NJ}{20} \quad \text{in 3D}.$$

Thus, to second-order for the 1D chain we obtain $E_G/N = -\frac{J}{4}(1+\gamma^2)$. Although perturbation theory assumes small γ, the result for $\gamma = 1$, $E_G/N = -\frac{J}{2}$, is nearer to the exact result of $-0.443147J$ (to 6 decimal places) than the zeroth-order value of $E_0/N = -\frac{J}{4}$, although it is still some distance away from the exact result.

Clearly, better accuracy can be obtained by including additional terms in the series, although this process becomes increasingly difficult to do analytically. Numerical computational techniques can, however, determine the series to high orders in γ.

Some of the earliest results were obtained by Bullock (1965) [8] and later ones by Singh (1989) [9]. For the antiferromagnetic ground state energy in 2D with $S = \frac{1}{2}$, Singh obtained the result

$$
\begin{aligned}
\frac{4E_G}{J} = {} & -2 - \tfrac{2}{3}\gamma^2 + 0.00370370\gamma^4 \\
& - 0.00632628\gamma^6 - 0.00330085\gamma^8 - 0.00124740\gamma^{10} + \ldots \quad .
\end{aligned}
$$

As noted above, even though the basic premise of perturbation theory is that $\mathcal{H}_1$ (and thus γ) is small, this series can give a reasonable estimate of the ground state energy even for the Heisenberg model with isotropic exchange, i.e. $\gamma = 1$.

Even with the use of powerful computers only a finite number of terms in the series can be calculated. A method of improving the results is to approximate the missing higher order terms using Padé approximants. This works well for many cases, although sometimes it is necessary to apply a transformation to avoid unphysical singularities. Using this approach (often also referred to as "series expansions" [10]), excellent results for the Heisenberg antiferromagnet have been achieved. Results of series expansions are discussed in the final chapter of this book.

Every numerical method has its own particular strengths and weaknesses. For example, for small systems, exact diagonalisations provide 'exact' results as the name suggests, and so are in some sense incontrovertible, which is a strong advantage of the method. However, the method is restricted to lattices with relatively small numbers of sites, $N \sim 40$ especially in 2D and 3D, even with the aid of high-performance computing. The DMRG method [11–14] provides essentially exact results for quasi-1D lattices, but has had only limited success in 2D.

As described above, another important method is the quantum Monte Carlo method (see, e.g., Refs. [4, 5]). In principle, this method gives results in which the accuracy is limited only by the amount of computational power available because the accuracy of the results increases in a statistically well-understood manner with the length of the Monte Carlo simulation. However, the method suffers from the 'sign-problem' and cannot be easily applied to 'frustrated' quantum spin systems.

Clearly there is a wide range of approximate techniques, each with advantages and disadvantages. In the following chapter we shall give a detailed account of a technique called the coupled cluster method (CCM) which has been applied to quantum spin systems only fairly recently. This is a technique related to that of cumulant series expansions and has important 'linked-cluster' properties. This method also gives results in the infinite-lattice limit ($N \to \infty$) from the outset, although it does not automatically provide an upper bound on the energy of the ground state or any other state.

References

1. Kalos, M.H.: Phys. Rev. **128**, 1791 (1962)
2. Ceperley, D.M., Kalos, M.H.: Quantum many-body problems. In: Binder, K. (ed.) Monte Carlo Methods in Statistical Physics, p. 145. Springer, Berlin (1979)
3. Schmidt, K., Kalos, M.H.: Few- and many-fermion problems. In: Binder, K. (ed.) Applications of the Monte Carlo Method in Statistical Physics, p. 125. Springer, Berlin (1984)
4. Runge, K.J.: Phys. Rev. B **45**, 12292 (1992)
5. Runge, K.J.: Phys. Rev. B **45**, 7229 (1992)
6. Sandvik, A.W.: Phys. Rev. B **56**, 11678 (1997)
7. Bishop, R.F., Farnell, D.J.J., Parkinson, J.B.: Phys. Rev. B **61**, 6775–6779 (2000)
8. Bullock, D.L.: Phys. Rev **137**, A1877–A1885 (1965)
9. Singh, R.R.P.: Phys. Rev. B **39**, 9760–9763 (1989)
10. Oitmaa, J., Hamer, C., Weihong, Z.: Series Expansion Methods for Strongly Interacting Lattice Models. Cambridge University Press, Cambridge Press, Cambridge (2006)
11. White, S.R.: Phys. Rev. Lett. **69**, 2863 (1992)
12. White, S.R., Huse, D.A.: Phys. Rev. B **48**, 3844 (1993)
13. White, S.R., Noack, R.M.: Phys. Rev. Lett. **68**, 3487 (1992)
14. White, S.R.: Phys. Rev. B **48**, 10345 (1993)

Chapter 10
The Coupled Cluster Method

Abstract Another powerful method of quantum many theory that obeys the linked-cluster theorem, and so provides results for the infinite lattice from the outset, is the Coupled Cluster Method (CCM). It has previously been applied to a wide range of quantum systems. We show here how it can be applied to quantum spin systems. The CCM formalism is described in detail. Crucial to understanding the CCM is the role of the model state upon which clusters of spin-raising operators act in order to form the basis states. Various approximation schemes may be used in order to calculate expectation values in practical calculations, examples being the LSUBm and SUBm approximation schemes. The LSUB2 and SUB2 approximations are presented in detail. Results for the LSUB2 approximation for the spin-half Heisenberg model on the linear chain are shown to be improved when compared to those of classical theory. We demonstrate how the CCM may be applied in order to study the ground- and excited-state properties of the anisotropic (XXZ) Heisenberg model on the square lattice. The CCM is shown is to provide an accurate and coherent picture for this model.

10.1 Introduction

In this chapter we consider another method that gives accurate results in the infinite lattice limit, especially for spin systems of two spatial dimensions, known as the coupled cluster method (CCM) [1]. The CCM is a well-known and widely applied method of quantum many-body theory. It allows us to calculate expectation values for the infinite lattice, which is a clear advantage to the method.

However, an aspect of the method is that one must often make an approximation within the bra- and ket-state wave functions, even though we obtain results in the infinite-lattice limit from the outset. The manner in which we construct these approximations is discussed below. Only lower orders of approximation than series expansions are possible, although the CCM contains many more diagrams than series expansions at "corresponding" levels of approximation. Furthermore, one does not necessarily obtain an upper bound on the ground-state energy using the CCM and no rules exist for extrapolation. Despite this, however, it is remarkable

Parkinson, J.B., Farnell, D.J.J.: *The Coupled Cluster Method*. Lect. Notes Phys. **816**, 109–134 (2010)
DOI 10.1007/978-3-642-13290-2_10

that the CCM still has been shown to provide consistently accurate results for a variety of quantum problems and for a wide range of expectation values (see final chapter). The CCM has been also applied with much success to quantum spin systems [2–10]. and it may be applied in the presence of even very strong frustration. Recent advances in the method have concentrated in applying it to high orders of approximation via computational methods [6–9].

10.2 The CCM Formalism

The first step in the coupled cluster method is to select an appropriate *model* state $|\Phi\rangle$. This state should be ideally be (i) simple and (ii) a reasonable starting approximation to the true ground state. The model state is assumed to act as a *vacuum* state so that a complete basis for any state can be constructed by operating on it by *creation* operators c_i^+. These creation operators act at a single lattice site i and a general state is obtained by operating on the model state with a linear combination of products of creation operators. We shall denote a product of creation operators, acting in general at several different sites, as C_I^+.

The central idea underpinning the CCM is to obtain better and better approximations to the true ground state $|\Psi\rangle$ by modifying the wave function in a systematic way, namely, by building in more and more of the true correlations with respect to the model state. Clearly there must be an operator P such that

$$|\Psi\rangle = P|\Phi\rangle. \tag{10.1}$$

However, rather than try to calculate P directly, we choose to introduce a new operator S where $P = e^S$ and attempt to calculate S instead. At first sight this seems an additional complication but the reason for it is as follows. By using this form it can be shown that any approximation we make by truncating S has the property that the equivalent diagrammatic perturbation theory approximation involves a summation only over *linked* diagrams. This is important since the Goldstone theorem states that only linked diagrams should be included if the calculated extensive property is to scale linearly with N, the size of the system.

It can be shown that the operator P and also the operator S consist of a sum of terms, where each term is itself a product of creation operators only, with respect to the model state. In the context of the Heisenberg model on a bipartite lattice such as the linear chain or square lattice, a creation operator is a spin-raising operator for sites where the spin is down or a spin lowering operator for sites where the spin is up. These creation operators are all mutually commuting. Hence, we write S as

$$S = \sum_{I \neq 0} S_I C_I^+, \tag{10.2}$$

where C_I^+ is a product of creation operators with associated (ket-state) correlation coefficient S_I.

For the exact ground state $|\Psi\rangle$

$$\mathcal{H}|\Psi\rangle = E_g|\Psi\rangle \quad \text{and} \quad \langle\tilde{\Psi}|\mathcal{H} = E_g\langle\tilde{\Psi}|, \tag{10.3}$$

where E_g is the exact ground state energy and $\langle\tilde{\Psi}|$ is the Hermitian conjugate of $|\Psi\rangle$. However, since we are writing

$$|\Psi\rangle = P|\Phi\rangle = e^S|\Phi\rangle, \tag{10.4}$$

and in most cases, S is an approximation to the true S, $|\Psi\rangle$ is an approximate ground state. It is obtained typically by truncating the otherwise infinite series of terms in S. Furthermore, we do not assume that the approximate $\langle\tilde{\Psi}|$ is the Hermitian conjugate of the approximate $|\Psi\rangle$. In fact, we shall construct $\langle\tilde{\Psi}|$ using a new auxiliary operator $\tilde{S}$ which is constructed from destruction operators only. $\tilde{S}$ is not the Hermitian conjugate of S and has to be obtained separately. The basic formulas are

$$|\Psi\rangle = e^S|\Phi\rangle \quad S = \sum_{I\neq 0} \mathcal{S}_I C_I^+, \tag{10.5}$$

$$\langle\tilde{\Psi}| = \langle\Phi|\tilde{S}e^{-S} \quad \tilde{S} = 1 + \sum_{I\neq 0} \tilde{\mathcal{S}}_I C_I^-. \tag{10.6}$$

This method of treating the bra state, using a linear operator $\tilde{S}$, is known as the normal coupled cluster method (NCCM). An alternative treatment is known as the Extended coupled cluster method (ECCM) in which the bra state is calculated using an exponentiated operator. The reader is referred to [11] for further details.

As mentioned earlier, the model or reference state $|\Phi\rangle$ plays the role of a vacuum state with respect to the $\{C_I^+\}$, i.e. their Hermitian conjugates $\{C_I^-\}$, have the property that $C_I^-|\Phi\rangle = 0, \forall I \neq 0$. We define that $C_0^+ \equiv C_0^- \equiv 1$ to be the identity operator. Furthermore, the set $\{C_I^+\}$ is complete and consists of all possible products of creation operators on multiple sites.

Also as mentioned earlier, the correlation operator S is composed entirely of the creation operators $\{C_I^+\}$, and these operators, acting on the model state, create other states in the relevant basis which are then mixed in to the model state to form an approximation to the 'true' ground state. Note that although the Hermiticity of the true ground state is lost, i.e. $\langle\tilde{\Psi}|^\dagger \neq |\Psi\rangle/\langle\Psi|\Psi\rangle$, we can still impose the normalisation conditions $\langle\tilde{\Psi}|\Psi\rangle = \langle\Phi|\Psi\rangle = \langle\Phi|\Phi\rangle \equiv 1$. The coefficients $\{\mathcal{S}_I\}$ and $\{\tilde{\mathcal{S}}_I\}$ are known as the ket- and bra-state *correlation coefficients*, respectively.

In general we need both $|\Psi\rangle$ and $\langle\tilde{\Psi}|$ (and hence need both $\{\mathcal{S}_I\}$ and $\{\tilde{\mathcal{S}}_I\}$) in order to find the ground-state expectation value of any operator, although the ground-state energy E_g is a special case which only requires knowledge of the $\{\mathcal{S}_I\}$.

For an arbitrary operator A the expectation value is given by,

$$\bar{A} \equiv \langle\tilde{\Psi}|A|\Psi\rangle = \langle\Phi|\tilde{S}e^{-S}Ae^S|\Phi\rangle = \bar{A}\left(\{\mathcal{S}_I, \tilde{\mathcal{S}}_I\}\right). \tag{10.7}$$

The similarity transform of the operator A, occurring here and denoted by $\hat{A}$, may be written as a series of nested commutators:

$$\hat{A} = e^{-S} A e^{S} = A + [A, S] + \frac{1}{2!}[[A, S], S] + \cdots \qquad (10.8)$$

In this expression each commutation reduces the number of destruction operators by one. As these commutations are nested within each other, the series will terminate at finite order provided that the operator A contains only a finite number of destruction operators. (There is no limit on the number of creation operators in A). N.B. we use the notation $\hat{}$ for a similarity transformed operator; we use the notation $\bar{}$ for an expectation value and the symbol $\tilde{}$ for the bra-state, its correlation operator $\tilde{S}$ and the corresponding coefficients $\{\tilde{S}_I\}$.

Using (10.3) and (10.4)

$$\mathcal{H} e^{S} |\Phi\rangle = E_g e^{S} |\Phi\rangle \qquad (10.9)$$

$$\therefore \quad e^{-S} \mathcal{H} e^{S} |\Phi\rangle = E_g |\Phi\rangle \qquad (10.10)$$

$$\text{i.e.} \quad \hat{\mathcal{H}} |\Phi\rangle = E_g |\Phi\rangle \qquad (10.11)$$

from which it immediately follows, using the normalisation $\langle \Phi | \Phi \rangle = 1$, that

$$E_g = \langle \Phi | \hat{\mathcal{H}} | \Phi \rangle \qquad (10.12)$$

From (10.10) $\quad C_I^- e^{-S} \mathcal{H} e^{S} |\Phi\rangle = E_g C_I^- |\Phi\rangle = 0$ since any destruction operator C_I^- acting on the vacuum state $|\Phi\rangle$ gives zero. Thus finally

$$\langle \Phi | C_I^- e^{-S} \mathcal{H} e^{S} |\Phi\rangle = \langle \Phi | C_I^- \hat{\mathcal{H}} |\Phi\rangle = 0. \qquad (10.13)$$

By choosing different C_I in Eq. (10.13) one obtains a coupled set of non-linear multinomial equations for the correlation coefficients $\{S_I\}$.

Also, using (10.3) and (10.6),

$$\langle \Phi | \tilde{S} e^{-S} \mathcal{H} = E_g \langle \Phi | \tilde{S} e^{-S}$$

$$\langle \Phi | \tilde{S} e^{-S} \mathcal{H} C_I^+ = E_g \langle \Phi | \tilde{S} e^{-S} C_I^+$$

$$\langle \Phi | \tilde{S} e^{-S} \mathcal{H} C_I^+ e^{S} |\Phi\rangle = E_g \langle \Phi | \tilde{S} e^{-S} C_I^+ e^{S} |\Phi\rangle$$

$$= \langle \Phi | \tilde{S} e^{-S} C_I^+ \mathcal{H} e^{S} |\Phi\rangle \quad \text{using (10.9)}$$

$$\therefore \quad \langle \Phi | \tilde{S} e^{-S} [\mathcal{H}, C_I^+] e^{S} |\Phi\rangle = 0 \qquad (10.14)$$

By choosing different C_I^+ in this equation one obtains a coupled set of *linear* multinomial equations for the correlation coefficients $\{S_I\}$.

For many purposes these three Eqs. (10.12), (10.13) and (10.14), together with (10.4), (10.5) and (10.6) form the essential core of the CCM method.

It is important to realise that, unlike conventional variational methods, this bi-variational formulation does *not* lead to an upper bound for E_g when the series in S and $\tilde{S}$ of Eqs. (10.5) and (10.6) are truncated. This is due to the lack of exact Hermiticity when such approximations are made.

The nested commutator expansion of the similarity-transformed Hamiltonian is given by

$$\hat{\mathcal{H}} \equiv e^{-S}\mathcal{H}e^{S} = \mathcal{H} + [\mathcal{H}, S] + \frac{1}{2!}[[\mathcal{H}, S], S] + \cdots . \tag{10.15}$$

This equation and the fact that all of the individual components of S in the sum in Eq. (10.5) commute with one another together imply that each element of S in Eq. (10.5) is linked directly to the Hamiltonian in each of the terms in Eq. (10.15). Equation (10.15) is therefore of *linked-cluster* type. As noted above, each of these equations is of finite length when expanded because the otherwise infinite series in Eq. (10.15) must always terminate at a finite order, provided only that each term in $\mathcal{H}$ contains a finite number of single-body destruction operators. Hence, the CCM parametrisation naturally leads to a workable scheme that can be carried out by hand for low orders of approximation or implemented computationally for higher orders of approximation. We stress that the similarity transformation lies at the heart of the CCM. This is in contrast to the *unitary* transformation that is at the heart of the standard variational formulation in which the bra state $\langle \tilde{\Psi} |$ is simply taken as the explicit Hermitian conjugate of $|\Psi\rangle$.

For the case of spin-lattice problems of the type considered here, the model state is usually the Néel state. Furthermore, it is useful to carry out a local rotation of the local spin axes at the 'up' sites so that these spins are all notionally pointing 'down'. This is purely a mathematical device; there is no physical rotation of the spins themselves but rather of the local axes we use to measure them. However, this does ensure that all the 'creation' operators with respect to the model state are now spin-raising operators of the form s_k^+ and the operators C_I^+ become products of these spin-raising operators only. This is very useful from a formal point of view because we treat all spins in exactly the same way regardless of whether they are on one sublattice or another.

We note that the CCM formalism would be exact if all possible multi-spin cluster correlations for S and $\tilde{S}$ were included. In any real application this is usually impossible to achieve. We remark again that it is therefore necessary to approximate the ground-state wave function. Indeed, we are able to construct approximation schemes within S and $\tilde{S}$ in which the number and/or type of clusters retained is restricted. The three most commonly employed schemes are:

(1) the **SUB***n* scheme, in which all correlations involving only *n* or fewer spins are retained, but no further restriction is made concerning their spatial separation on the lattice;
(2) the **SUB***n-m* sub-approximation, in which all SUB*n* correlations spanning a range of no more than *m* adjacent lattice sites are retained; and

(3) the localised **LSUB**m scheme, in which all multi-spin correlations over all distinct locales on the lattice defined by m or fewer contiguous sites are retained. The problem of solving for these types of approximation schemes using analytical and computational approaches is discussed below.

All of these approximation schemes follow the 'rule' that defines a true quantum many-body theory, namely, that we may increase the level of approximation in a systemic and well-controlled manner. Furthermore, we can also attempt to extrapolation our 'raw' SUBn, SUBn-m, and LSUBm results in the limits $n, m \to \infty$. However, by contrast to exact diagonalisations and quantum Monte Carlo in which finite-sized lattice results are extrapolated in the infinite lattice limit using well-defined extrapolation procedures, no such equivalent extrapolation schemes exist as yet for the CCM. We are therefore forced to use 'heuristic' or 'ad hoc' schemes in order to extrapolate our results. An example of an 'heuristic' extrapolation scheme of LSUBm data for the ground state energy is a polynomial fit given by $y = a + bm^{-2} + cm^{-4}$. A similar polynomial fit for the sublattice magnetisation is $y = a + bm^{-1} + cm^{-2}$, although a power-law fit, i.e., $y = a + bm^{-\nu}$, is also often used in this case. The lack of an upper bound on the ground-state energy (due to the fact that bra and ket states are not explicitly constrained to be Hermitian conjugates) is not the biggest problem in practice. Indeed, the CCM often does provide an upper bound for those cases in which the model state is believed to be a reasonable "starting point." In fact, the biggest limitation of the CCM in practice is the lack of concrete "rules" for extrapolation of LSUBm results and the (sometimes) rather small number of LSUBm results to extrapolate with – even with intensive computer methods. However, it is remarkable that, despite these potential limitations, the CCM often does provide accurate results compared to results of exact studies and the best of other approximate methods. This is demonstrated later on in this chapter and also in the next.

The NCCM may also be used to investigate excited states. In order to do this we introduce a third *excited-state* operator X^e (in addition to the CCM ground-state ket- and bra-state operators, S and $\tilde{S}$). This operator is again a linear combination of the C_I^+ with associated coefficients $\{\mathcal{X}_I^e\}$

$$X^e = \sum_{I \neq 0} \mathcal{X}_I^e C_I^+. \tag{10.16}$$

We see readily that X^e commutes with S as it contains only the set $\{C_I^+\}$ of multi-spin creation operators. However, the specific clusters used in the set $\{C_I^+\}$ may differ from those used in the ground-state parametrisation in Eqs. (10.5) and (10.6) if the excited state has different quantum numbers than the ground state. An excited-state wave function, $|\Psi_e\rangle$, is determined by applying X^e to the ket-state wave function of Eq. (10.5) such that

$$|\Psi_e\rangle = X^e \, e^S |\Phi\rangle. \tag{10.17}$$

The energy E_e of the excited state is given by the Schrödinger equation, where

$$\mathcal{H}|\Psi_e\rangle = E_e|\Psi_e\rangle. \tag{10.18}$$

We may apply X^e to the CCM ground-state Schrödinger equation such that $X^e\mathcal{H}|\Psi\rangle = E_g X^e|\Psi\rangle$. This expression, in turn, leads to

$$E_e|\Psi_e\rangle - E_g X^e|\Psi\rangle \equiv \mathcal{H}|\Psi_e\rangle - X^e\mathcal{H}|\Psi\rangle$$

$$\Rightarrow E_e X^e e^S|\Phi\rangle - E_g X^e e^S|\Phi\rangle = \mathcal{H}X^e e^S|\Phi\rangle - X^e\mathcal{H}e^S|\Phi\rangle$$

$$\text{or } \varepsilon_e X^e|\Phi\rangle = e^{-S}[\mathcal{H}, X^e]e^S|\Phi\rangle, \tag{10.19}$$

where $\varepsilon_e \equiv E_e - E_g$ is the excitation energy and we note that $[X^e, e^S] = 0$. Equation (10.17) implies that $\langle\Phi|\Psi_e\rangle = 0$. Thus, we find by applying $\langle\Phi|C_I^-$ to Eq. (10.19) that,

$$\varepsilon_e \mathcal{X}_I^e = \langle\Phi|C_I^- e^{-S}[\mathcal{H}, X^e]e^S|\Phi\rangle, \forall I \neq 0, \tag{10.20}$$

which is a generalised set of eigenvalue equations with eigenvalues ε_e and corresponding eigenvectors $\mathcal{X}_I^e$, for the excited states.

Again, we note that it is sometimes possible to solve these sets of equations by hand for low orders of approximation. However, it rapidly becomes clear that analytical determination of the CCM equations for higher orders of approximation is impractical and it is therefore necessary to employ computer algebraic techniques both to determine and to solve the equations. Once the bra- and ket-state equations have been determined they are readily solved using standard techniques for the solution of coupled polynomial equations (e.g., the Newton-Raphson method). The excited-state eigenvalue equations may be also determined and solved computationally thereafter. A full description of the details in applying the CCM to high orders of approximation is given for the ground state in Bishop et al. [7]. We have seen above that we are able to increase the level of approximation for the the SUBn, SUBm-m and LSUBm approximation schemes in in the ground state in a systematic way. This holds true also for the excited states. Thus, excited-state energies may again be extrapolated to the 'exact limit' $n, m \to \infty$ using a variety of 'heuristic' approaches.

10.3 The XXZ-Model

In this chapter we shall use lower case s^z, etc., for spin operators to avoid confusion with the capital S and $\tilde{S}$ which are the CCM correlation operators.

The spin-half XXZ antiferromagnetic model on the square lattice has a Hamiltonian given by

$$\mathcal{H} = \sum_{\langle i,j\rangle}[s_i^x s_j^x + s_i^y s_j^y + \Delta s_i^z s_j^z] = \frac{1}{2}\sum_{\langle i,j\rangle}[s_i^+ s_j^- + s_i^- s_j^+ + 2\Delta s_i^z s_j^z], \tag{10.21}$$

where the sum on $\langle i, j \rangle$ counts all nearest-neighbour pairs once. The Néel state, with all 'down' spins on one sublattice and all 'up' spins on the other, is the ground state in the (trivial) Ising limit $\Delta \to \infty$. As Δ decreases the ground state remains Néel-like until a phase transition occurs at (or near to) $\Delta = 1$. By Néel-like we mean that there is a substantial positive expectation value of $\langle s^z \rangle$ for spins on one sublattice and an equal and opposite expectation value for those spins on the other. Even at $\Delta = 1$ (i.e., the Heisenberg model), approximately 61% to 62% of the classical ordering remains in the quantum system. For $-1 < \Delta < 1$ the ground state is co-planar, with zero expectation value for $\langle s^z \rangle$; the atoms are aligned in the xy-plane. For $\Delta < -1$ the system is ferromagnetic and the exact ground state has all atoms aligned in the z-direction.

This Néel state is the obvious choice for $|\Phi\rangle$ in the region $\Delta > 1$ which is known as the Néel-like region. It is convenient to carry out a transformation of the local spin axes at each site on one of the sublattices by performing a rotation of the up-pointing spins by $180°$ about the y-axis, such that

$$x \to -x, \qquad y \to y, \qquad z \to -z \qquad (10.22)$$

and the spin components transform as

$$s^x \to -s^x, \qquad s^y \to s^y, \qquad s^z \to -s^z \qquad (10.23)$$

and so

$$s^+ = s^x + is^y \to -s^- \quad \text{and} \quad s^- = s^x - is^y \to -s^+. \qquad (10.24)$$

The effect of this transformation is that every spin, whichever sublattice it is on, is now (notionally) pointing 'down' in the Néel state, i.e. with $s^z = -\frac{1}{2}$. This makes the process of determining the CCM equations easier as each site may now be treated equally. The Hamiltonian of Eq. (10.21) in these local coordinates now becomes

$$\mathcal{H} = -\frac{1}{2} \sum_{\langle i,j \rangle} [s_i^+ s_j^+ + s_i^- s_j^- + 2\Delta s_i^z s_j^z]. \qquad (10.25)$$

The transformation is canonical and does not alter the commutation relations between the spin operators on a given site. (All spin operators referring to different sites commute). Furthermore, it does not alter the values of the ground-state expectation values or the the excited state energies or spectra. Apart from this 'down' spin state with $s^z = -\frac{1}{2}$, there is only one 'other' state at each site in this new basis, namely, the 'up' state with $s^z = +\frac{1}{2}$. The creation operators used in the CCM ket-state correlation operator S are now clearly always the spin-raising operators s^+. A C_I^+ is a product of these spin-raising operators acting at different sites. (Note that the creation operator cannot act more than once at a given site for a spin-$\frac{1}{2}$ atom, although this restriction would not apply for $s > \frac{1}{2}$.) An example for $s = \frac{1}{2}$ might be $C_I^+ = s_i^+ s_j^+ s_k^+ s_l^+$ in which i, j, k, l are different sites on the lattice.

The transformation is purely a mathematical device, there is no physical rotation of the spins. When considering which other states can be mixed in with the Néel state to form an approximation to the true ground state we must take into account any physical properties we know it must satisfy. In particular, for the Hamiltonian Eq. (10.21) the total z-component of the spins $s_T^z = \sum_i s_i^z$ is a conserved quantity. Since the Néel state has $s_T^z = 0$ we can only mix in states with the same S_T^z and this means that C_I^+ must contain an equal number of spin reversals on each sublattice. Clearly there must be an even number of spin-raising operators, half from each sublattice.

The results presented below are based on the non-localised SUB2 approximation scheme and the localised LSUBm scheme. In the latter we include all *fundamental configurations*, $C_I^+ = s_{k_1}^+, s_{k_2}^+, \cdots s_{k_n}^+$, where the number of contiguous sites is $\leq m$. Fundamental configurations are those which are distinct under the point and space group symmetries of both the lattice and the Hamiltonian. The numbers, N_F and N_{F_e}, of such fundamental configurations for the ground and excited states, respectively, are also further restricted by the use of conservation laws, in particular conservation of s_T^z, as mentioned above. As well as $s_T^z = 0$ for the ground state we have $s_T^z = \pm 1$ for the elementary excited states.

10.3.1 The LSUB2 Approximation for the Spin-Half, Square-Lattice XXZ-Model for the z-Aligned Model State

In the LSUB2 approximation we allow two creation operators and they must be on nearest neighbour sites. The only possible C_I^+, other than C_0^+, are terms of the form $s_l^+ s_{l+\rho_1}^+$ where l is any lattice site and ρ_1 is a vector connecting nearest neighbours. The form of the S operator is thus

$$ S = \frac{b_1}{2} \sum_l^N \sum_{\rho_1} s_l^+ s_{l+\rho_1}^+, \tag{10.26} $$

where l runs over all lattice sites and ρ_1 runs over all nearest-neighbour sites to l. Note that b_1 is the sole ket-state correlation coefficient in the LSUB2 approximation scheme.

We now calculate the similarity transforms of the operators in the Hamiltonian, $e^{-S} s_k^\alpha e^S$ for $\alpha = z, +, -$. The commutation relations for the spin operators are given by $[s_l^\pm, s_k^z] = \mp s_k^\pm \delta_{l,k}$ and $[s_l^+, s_k^-] = 2s_k^z \delta_{l,k}$. Furthermore, the similarity transform may be expanded as a series of nested commutators, given by Eq. (10.8). Hence, we obtain the following explicit forms for these similarity transformed operators

$$ \hat{s}_i^+ = s_i^+ $$
$$ \hat{s}_i^z = s_i^z + b_1 \sum_{\rho_1} s_i^+ s_{i+\rho_1}^+ \tag{10.27} $$
$$ \hat{s}_i^- = s_i^- - 2b_1 \sum_{\rho_1} s_i^z s_{i+\rho_1}^+ - b_1^2 \sum_{\rho_1,\rho_2} s_i^+ s_{i+\rho_1}^+ s_{i+\rho_2}^+. $$

In each case the otherwise infinite series of operators in the expansion of the similarity transform has terminated to finite order. All of Eq. (10.27) are valid for arbitrary spin, but since we are considering only spin-half systems here for which $(s_i^+)^2|\Phi\rangle = 0$ for *any* lattice site, the term in the third equation in the summations over ρ_1 and ρ_2 for which $\rho_1 = \rho_2$ will be zero. Clearly the similarity transformed version of the Hamiltonian is

$$\hat{\mathcal{H}} = -\frac{1}{2}\sum_{\langle i,j\rangle}[\hat{s}_i^+\hat{s}_j^+ + \hat{s}_i^-\hat{s}_j^- + 2\Delta\hat{s}_i^z\hat{s}_j^z]. \tag{10.28}$$

Note that the sum over $\langle i, j\rangle$ is equivalent to a sum over all sites i and over all nearest neighbours ρ_0, together with a factor of $\frac{1}{2}$ to avoid overcounting, so

$$\hat{\mathcal{H}} = -\frac{1}{4}\sum_{i}\sum_{\rho_0}[\hat{s}_i^+\hat{s}_{i+\rho_0}^+ + \hat{s}_i^-\hat{s}_{i+\rho_0}^- + 2\Delta\hat{s}_i^z\hat{s}_{i+\rho_0}^z]. \tag{10.29}$$

Substituting the expressions for the spin operators in Eq. (10.27) into the above expression, gives

$$\hat{\mathcal{H}} = -\frac{1}{4}\sum_{i}\sum_{\rho_0}[s_i^+s_{i+\rho_0}^+ + \{s_i^- - 2b_1\sum_{\rho_1}s_i^z s_{i+\rho_1}^+ - b_1^2\sum_{\rho_1,\rho_2}s_i^+ s_{i+\rho_1}^+ s_{i+\rho_2}^+\} \times$$
$$\{s_{i+\rho_0}^- - 2b_1\sum_{\rho_3}s_{i+\rho_0}^z s_{i+\rho_0+\rho_3}^+ - b_1^2\sum_{\rho_3,\rho_4}s_{i+\rho_0}^+ s_{i+\rho_0+\rho_3}^+ s_{i+\rho_0+\rho_4}^+\} +$$
$$2\Delta\{s_i^z + b_1\sum_{\rho_1}s_i^+ s_{i+\rho_1}^+\} \times \{s_{i+\rho_0}^z + b_1\sum_{\rho_2}s_{i+\rho_0}^+ s_{i+\rho_0+\rho_2}^+\}]. \tag{10.30}$$

When this $\hat{\mathcal{H}}$ is now used in Eq. (10.13) with $C_I^- = s_m^- s_{m+\rho}^-$ to determine the coefficient b_1, only terms with net two spin-raising operators are needed. When used in Eq. (10.12) for the ground state energy E_g only terms with net zero spin-raising operators are needed. In addition, when calculating the bra state coefficient using Eq. (10.14) terms with net two lowering operators will be needed. Keeping only the terms in Eq. (10.29) with these forms leads to the following simplified expression

$$\hat{\mathcal{H}} \approx \hat{\mathcal{H}}_{-2} + \hat{\mathcal{H}}_0 + \hat{\mathcal{H}}_2$$

where

$$\hat{\mathcal{H}}_{-2} = -\frac{1}{4}\sum_{i}\sum_{\rho_0} s_i^- s_{i+\rho_0}^-$$

is the part with net two spin lowering operators,

$$\hat{\mathcal{H}}_0 = -\frac{1}{4}\sum_i \sum_{\rho_0}\left[-2b_1\sum_{\rho_3} s_i^- s_{i+\rho_0}^z s_{i+\rho_0+\rho_3}^+ - 2b_1\sum_{\rho_1} s_i^z s_{i+\rho_1}^+ s_{i+\rho_0}^- + 2\Delta s_i^z s_{i+\rho_0}^z\right]$$

$$(10.31)$$

is the part with net zero spin-raising operators, and

$$\hat{\mathcal{H}}_2 = -\frac{1}{4}\sum_i \sum_{\rho_0}\left[s_i^+ s_{i+\rho_0}^+ + 4b_1^2 \sum_{\rho_1,\rho_3} s_i^z s_{i+\rho_1}^+ s_{i+\rho_0}^z s_{i+\rho_0+\rho_3}^+ \right.$$

$$- b_1^2 \sum_{\rho_3,\rho_4} s_i^- s_{i+\rho_0}^+ s_{i+\rho_0+\rho_3}^+ s_{i+\rho_0+\rho_4}^+ - b_1^2 \sum_{\rho_1,\rho_2} s_i^+ s_{i+\rho_1}^+ s_{i+\rho_2}^+ s_{i+\rho_0}^-$$

$$\left. + 2\Delta b_1 \sum_{\rho_2} s_i^z s_{i+\rho_0}^+ s_{i+\rho_0+\rho_2}^+ + 2\Delta b_1 \sum_{\rho_1} s_i^+ s_{i+\rho_1}^+ s_{i+\rho_0}^z \right]$$

$$(10.32)$$

is the part with net two spin-raising operators.

First consider Eq. (10.12) for the ground-state energy

$$E_g = \langle\Phi|\hat{\mathcal{H}}|\Phi\rangle = \langle\Phi|\hat{\mathcal{H}}_0|\Phi\rangle.$$

Using the commutator $[s_i^-, s_{i+\rho_0+\rho_3}^+] = -2s_i^z \delta_{i,i+\rho_0+\rho_3}$ gives

$$\hat{\mathcal{H}}_0 = -\frac{1}{4}\sum_i \sum_{\rho_0}\left[-2b_1\sum_{\rho_3} s_{i+\rho_0}^z s_{i+\rho_0+\rho_3}^+ s_i^- + 4b_1 s_{i+\rho_0}^z s_i^z \right.$$

$$\left. -2b_1 \sum_{\rho_1} s_i^z s_{i+\rho_1}^+ s_{i+\rho_0}^- + 2\Delta s_i^z s_{i+\rho_0}^z\right],$$

$$(10.33)$$

and when this acts on $|\Phi\rangle$ the terms with a spin lowering operator on the right will give zero. Hence

$$\hat{\mathcal{H}}_0|\Phi\rangle = -\frac{1}{4}\sum_i \sum_{\rho_0}\left[4b_1 s_i^z s_{i+\rho_0}^z + 2\Delta s_i^z s_{i+\rho_0}^z\right]|\Phi\rangle$$

$$(10.34)$$

$$= -\frac{1}{16}\sum_i \sum_{\rho_0}(4b_1 + 2\Delta)|\Phi\rangle.$$

$$(10.35)$$

since $s_i^z|\Phi\rangle = -\frac{1}{2}|\Phi\rangle$ for all i.

Using Eq. (10.12), the ground-state energy is

$$\frac{E_g}{N} = -\frac{n}{8}(\Delta + 2b_1).$$

$$(10.36)$$

where n is the number of nearest neighbours.

Equation (10.36) shows that the ground-state energy is size-extensive (i.e., it scales linearly with N), as required by the Goldstone theorem which is obeyed by the NCCM. In fact it is easy to show that *any* other non-trivial choice for S, not just the LSUB2 approximation, will always yield expression (10.36) for the ground-state energy, although the calculation of b_1 will be different. The task is therefore to find b_1. If we could include all possible spin correlations in S then we would obtain an exact result for b_1 and hence the ground-state energy. This is of course impossible except for trivial cases so there will normally need to be an approximation like the ones described here.

In this LSUB2 approximation b_1 is the only non-zero coefficient in S and it is determined using $\hat{\mathcal{H}}_2$ in Eq. (10.13), with $C_I^- = s_m^- s_{m+\rho}^-$

$$\langle \Phi | C_I^- \hat{\mathcal{H}} | \Phi \rangle = \langle \Phi | C_I^- \hat{\mathcal{H}}_2 | \Phi \rangle = 0. \tag{10.37}$$

since C_I^- has two lowering operators. The details of the calculation are given in the appendix where it is shown that Eq. (10.37) yields

$$(n+1)b_1^2 + 2(n-1)\Delta b_1 - 1 = 0. \tag{10.38}$$

Equations (10.36) and (10.38) are the basic equations in the LSUB2 approximation for the linear chain with $n = 2$, the square lattice with $n = 4$ and the cubic lattices. Similar results can be obtained for any other bipartite lattice with nearest-neighbour interactions.

For the linear chain these equations become

$$\frac{E_g}{N} = -\frac{1}{4}(\Delta + 2b_1) \quad \text{with} \quad 3b_1^2 + 2\Delta b_1 - 1 = 0, \tag{10.39}$$

so that

$$b_1 = \frac{1}{3}(\sqrt{\Delta^2 + 3} - \Delta) \quad \text{and} \quad \frac{E_g}{N} = -\frac{\Delta}{12} - \frac{\sqrt{\Delta^2 + 3}}{6}. \tag{10.40}$$

This gives a value for ground-state energy the isotropic Heisenberg model ($\Delta = 1$) of $\frac{E_g}{N} = -\frac{5}{12} (\equiv -0.416667)$ (to 6 decimal places), which compares to the exact result of $\frac{E_g}{N} = -0.443147 J$ (again to 6 decimal places). This is an improvement on energy of the (classical) model state, which is $\frac{E_g}{N} = -\frac{1}{4}$.

However, we shall consider only the square lattice with $n = 4$ from now on, for which

$$\frac{E_g}{N} = -\frac{1}{2}(\Delta + 2b_1) \quad \text{with} \quad 5b_1^2 + 6\Delta b_1 - 1 = 0, \tag{10.41}$$

so that

$$b_1 = \frac{1}{5}(\sqrt{9\Delta^2 + 5} - 3\Delta) \quad \text{and} \quad \frac{E_g}{N} = \frac{\Delta}{10} - \frac{1}{5}\sqrt{9\Delta^2 + 5} . \quad (10.42)$$

This expression gives the correct result in the Ising limit $\Delta \to \infty$. These results for the ground-state energy as a function of Δ in the LSUB2 approximation are included in both Figs. 10.1 and 10.2.

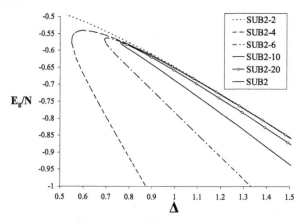

Fig. 10.1 CCM SUB2-m and SUB2 results using the z-aligned Néel model state for the ground-state energy of the spin-half square-lattice XXZ-Model. (Note that SUB2-2 and LSUB2 are equivalent approximations)

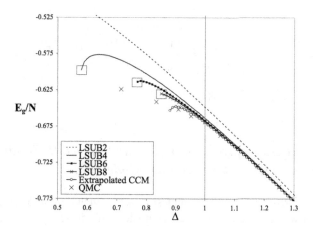

Fig. 10.2 CCM LSUBm results for the ground-state energy of the spin-half square-lattice XXZ-Model compared to quantum Monte Carlo results of Barnes et al. [12]

A similar calculation, based on Eq. (10.14) although not using it directly, gives the following equation for $\tilde{b}_1$. Again details are given in the appendix.

$$\tilde{b}_1[(n + 1)2b_1 + 2(n - 1)\Delta] - 1 = 0. \quad (10.43)$$

For the square lattice with $n = 4$ this becomes

$$10b_1\tilde{b}_1 + 6\Delta\tilde{b}_1 - 1 = 0, \tag{10.44}$$

which gives $\tilde{b}_1 = \frac{1}{2}(9\Delta^2 + 5)^{-1/2}$.

Finally, we note that once the values for the bra- and ket-state correlation coefficients have been determined (at a given level of approximation) then we can also evaluate various expectation values. An important example is the sublattice magnetisation given by

$$M \equiv \frac{2}{N}\langle\tilde{\Psi}|\sum_{i}^{N}(-1)^{i}s_{i}^{z}|\Psi\rangle, \tag{10.45}$$

in terms of the original unrotated spin coordinates. After rotation of the local spin axes

$$M = -\frac{2}{N}\langle\tilde{\Psi}|\sum_{i}^{N}s_{i}^{z}|\Psi\rangle = -\frac{2}{N}\langle\Phi|\tilde{S}e^{-S}\left(\sum_{i}^{N}s_{i}^{z}\right)e^{S}|\Phi\rangle, \tag{10.46}$$

in terms of the 'rotated' spin coordinates. For the square lattice in the LSUB2 approximation this is given by

$$M_{\text{LSUB2}} = 1 - 8b_1\tilde{b}_1,$$
$$= \frac{1}{5}\left[1 + \frac{12\Delta}{\sqrt{9\Delta^2 + 5}}\right]. \tag{10.47}$$

This result is shown in Fig. 10.3.

10.3.2 The SUB2 Approximation for the Spin-Half, Square-Lattice XXZ-Model of the z-Aligned Model State

The LSUB2 approximation is the simplest possible, including in S terms with just two spin flips which have to be on adjacent sites. Of course the exponentiation of the S operator, e^S, results in multiple applications of this and so the approximate ground state calculated in the previous section includes contributions from states with arbitrarily large numbers of flips at widely separated sites.

The SUB2 approximation is a generalisation of this in which all possible two-spin-flip terms are included in S. These can now be at any two sites although, of course one must be on one sublattice and the other on the other sublattice. The SUB2 ket-state operator S is given by

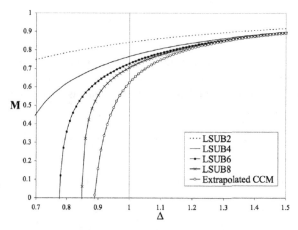

Fig. 10.3 CCM LSUBm results using the z-aligned Néel model states for the sublattice magneti-sation of the spin-half square-lattice XXZ-Model

$$S = 1/2 \sum_i^N \sum_r b_r s_i^+ s_{i+r}^+, \tag{10.48}$$

where the index i runs over all sites on the lattice chain. The index r runs over all lattice vectors which connect one sublattice to the other and b_r is the corresponding SUB2 ket-state correlation coefficient for this vector. The similarity transformed versions of the operators are

$$
\begin{aligned}
\tilde{s}_l^+ &= s_l^+ \\
\tilde{s}_l^z &= s_l^z + \sum_r b_r s_l^+ s_{l+r}^+ \\
\tilde{s}_l^- &= s_l^- - 2\sum_r b_r s_l^z s_{l+r}^+ - \underbrace{\sum_{r_1, r_2} b_{r_1} b_{r_2} s_l^+ s_{l+r_1}^+ s_{l+r_2}^+}_{r_1 \neq r_2}.
\end{aligned}
\tag{10.49}
$$

Again we note that $(s_i^+)^2 |\Phi\rangle = 0$ for *any* lattice site (which is true only for spin-half systems). Hence, we see that $r_1 \neq r_2$ in the above equations. We now substitute these expressions into Eq. (10.25) in order to obtain $\tilde{\mathcal{H}}$, and we then use Eq. (10.13) in order to obtain the following equation for the correlation coefficients $\{b_r\}$

$$\sum_\rho \left\{ (1 + 2\Delta b_1 + 2b_1^2)\delta_{\rho,r} - 2(\Delta + 2b_1)b_r + \sum_s b_{r-s-\rho_1} b_{s+\rho+\rho_1} \right\} = 0, \tag{10.50}$$

where ρ runs over all nearest-neighbour vectors on the square lattice and ρ_1 is any one of them. Equation (10.50) may now be solved using a sublattice Fourier transform, given by

$$\Gamma(q) = \sum_r e^{i\mathbf{r}\cdot\mathbf{q}} b_r, \tag{10.51}$$

where the sum over r is a sum over the position vectors of *one* sublattice only. A convenient way to achieve this for the two-dimensional square lattice is to require that $r_x + r_y$ is an odd integer number. This expression has an inverse given by

$$b_r = \int_{-\pi}^{\pi} \int_{-\pi}^{\pi} \frac{dq_y dq_x}{(2\pi)^2} \cos(r_x q_x) \cos(r_y q_y) \Gamma(q). \tag{10.52}$$

The SUB2 Eqs. (10.50) and (10.51) lead to the following expression [5] for $\Gamma(q)$

$$\Gamma(q) = \frac{K}{\gamma(q)} \left[1 \pm \sqrt{1 - k^2 \gamma^2(q)} \right], \tag{10.53}$$

$$K = \Delta + 2b_1, \quad k^2 = (1 + 2\Delta b_1 + 2b_1^2)/K^2 \text{ and}$$

$$\gamma(q) = \frac{1}{n} \sum_{\rho} e^{i\rho\cdot\mathbf{q}} \left(\equiv \frac{1}{2} \{\cos(q_x) + \cos(q_y)\} \text{ for the square lattice} \right) \tag{10.54}$$

(Note that we choose the negative solution in Eq. (10.53) such that the result is correct in the trivial limit $\Delta \to \infty$.) Equations (10.52), (10.53) and (10.54) lead to a self-consistency requirement on the variable b_1 and they may be solved iteratively at a given value of Δ. Indeed, we know that all correlation coefficients must tend to zero (namely, for SUB2: $b_r \to 0$, $\forall r$) as $\Delta \to \infty$ and we *track* this solution for large Δ by reducing Δ in small successive steps. We find that the discriminant in Eq. (10.53) becomes negative at a *critical point*, $\Delta_c \approx 0.7985$. This is an indication that the CCM critical point corresponds to a quantum phase transition in the system, although in this simple approximation scheme it is some way from the known phase transition at $\Delta = 1$.

We may also solve the SUB2-m equations directly using computational techniques. (Note that the SUB2-2 and LSUB2 approximations are equivalent.) Indeed, we study the limit points of these coupled non-linear equations may be obtained with respect to Δ. We again track our solution from the limit $\Delta \to \infty$ to (and beyond) the limit point and Fig. 10.1 shows our results. In particular, we note that we have two distinct branches, although only the upper branch is a 'physical' solution. We have already remarked that the CCM does not necessarily always provide an upper bound on the ground-state energy, although this is often the case for the 'physical' solution. An example of this is seen by the 'unphysical' lower branch in Fig. 10.1. However, the solution of the CCM equations will often naturally converge to the physical solution, provided we have a reasonable starting point for the CCM correlation coefficients and our model state is also reasonable. However, by tracking from a point at which we are sure of (in this case, from the limit $\Delta \to \infty$), we ensure that our solution is indeed the correct one. This approach is also used for LSUBm approximations.

We see in Fig. 10.1 that the two branches collapse onto the same line (namely, that of the full SUB2 solution) as we increase the level of SUB2-m approximation with respect to m. Indeed, we may plot the positions of the SUB2-m limit points against $1/m^2$ and we note that these data points are found to be both highly linear and they tend to the critical value, Δ_c, of the full SUB2 equations in the limit $m \to \infty$. Again, we note that the LSUBm and SUBm-m a approximations also show similar branches (namely, one 'physical' and one 'unphysical' branch) which appear to converge as one increases the magnitude of the truncation index, m, although results for the 'unphysical' branches are not presented here for the LSUBm approximation. This is a strong indication that our LSUBm and SUBm-m critical points are also reflections of phase transitions in the real system. We expect that our extrapolated LSUBm and SUBm-m critical points should tend to the exact solution.

10.3.3 High-Order CCM Calculations Using a Computational Approach

We now consider the localised LSUBm and SUBm-m approximation schemes for larger values of m than $m = 2$. Recall that LSUBm allows all possible terms in S in which spin flips all occur within a *locale* of size m. For spin-half the maximum number of spin-raising operators in one term is m, but for general spin quantum number, s, it is $2sm$. SUBm-n is a scheme in which one allows a maximum of m spin-raising operators in any one term and restricts them to all lie within a locale of size n. This locale is defined by those configurations that contain m contiguous sites. For spin-half systems, SUBm-m is the same as LSUBm and this is the only type considered here. These schemes are more complicated and cannot usually be treated analytically as we were able to do for LSUB2 and SUB2. Consequently, computational techniques are used both to determine the CCM equations and then to solve them numerically.

There are two methods of doing this. Firstly, one may use computer algebraic methods to calculate the similarity transformed versions of the individual spin operators and hence the similarity transformed version of the Hamiltonian, which may involve further commutations of the spin operators. This approach has the advantage of flexibility and can be applied to any Hamiltonian in principle. Often, however, this method is somewhat cumbersome and slow.

A second method is to first cast the CCM ket-state correlation operator into a form given by

$$S = \sum_{i_1}^{N} \mathcal{S}_{i_1} s_{i_1}^+ + \sum_{i_1,i_2} \mathcal{S}_{i_1,i_2} s_{i_1}^+ s_{i_2}^+ + \cdots \tag{10.55}$$

with respect to a model state in which all spins point in the downwards z-direction. Here the $\mathcal{S}_{i_1,\cdots,i_l}$ represent the CCM ket-state correlation coefficients as in Eq. (10.5). We now define new operators given by

$$F_k = \sum_l \sum_{i_2,\cdots,i_l} l\mathcal{S}_{k,i_2,\cdots,i_l} s_{i_2}^+ \cdots s_{i_l}^+$$

$$G_{k,m} = \sum_{l>1} \sum_{i_3,\cdots,i_l} l(l-1)\mathcal{S}_{k,m,i_3,\cdots,i_l} s_{i_3}^+ \cdots s_{i_l}^+ \qquad (10.56)$$

for the spin-half quantum spin systems. (For $s > 1/2$ additional terms are needed). For the spin-half system the similarity transformed operators can be written

$$\tilde{s}_k^+ = s_k^+$$
$$\tilde{s}_k^z = s_k^z + F_k s_k^+$$
$$\tilde{s}_k^- = s_k^- - 2F_k s^z - (F_k)^2 s_k^+. \qquad (10.57)$$

These expressions can now be substituted analytically into the (similarity transformed) Hamiltonian and the commutations evaluated by hand. The Hamiltonian is then written in terms of these new operators of Eq. (10.56), which are themselves made up purely of spin-raising operators.

This second method requires more direct effort in setting up the Hamiltonian in terms of these new operators, compared to the first method in which computer algebraic techniques are used to take care of this aspect. However, once this is accomplished, the problem of finding the ket-state equations reduces to pattern matching of our target fundamental configurations to those terms in the Hamiltonian. This form is well suited to a computational implementation because no further commutations or re-ordering of terms in the Hamiltonian is necessary. The bra-state equations may also be directly determined once the ground-state energy and CCM ket-state equations have been determined.

Results for the ground-state energy of the spin-half square-lattice XXZ-Model are shown in Fig. 10.2 and for the spin-half Heisenberg model ($\Delta = 1$) in Table 10.1. We note that good correspondence with the results of quantum Monte Carlo (QMC) [12] are observed. The extrapolated value for the CCM ground-state energy of $E_g/N = -0.6696$ compares well with results of QMC [13] that give $E_g/N = -0.669437(5)$. We see clearly from Fig. 10.2 that the results based on the (z-aligned) model state rapidly converge with increasing level of LSUBm approximation in the region $\Delta \geq 1$. CCM results compare well to results of QMC [12] for $\Delta \geq 1$ based on this model state. Results for the sublattice magnetisation using the LSUBm approximation are shown in Fig. 10.3 and for the spin-half Heisenberg model ($\Delta = 1$) in Table 10.1. We see again that LSUBm results converge rapidly with increasing level of approximation in the region $\Delta \geq 1$. We see that the extrapolated CCM result of $M = 0.614$ again compares well to QMC results [13] of $M = 0.6140(6)$. Results for the CCM critical points are also shown in Table 10.1. We see that the values for the critical points, Δ_c, extrapolate [7] to a value close to $\Delta = 1$, at (or near to) which point a quantum phase transition is believed to occur.

Table 10.1 CCM results [7] for the isotropic ($\Delta = 1$) spin-half square-lattice Heisenberg antiferromagnet compared to results of other methods. The numbers of fundamental configurations in the ground-state and excited-state CCM wave functions for the z-aligned Néel model state are given by N_f^z and $N_{f_e}^z$, respectively. Results for the critical points of the z-aligned Néel model state are indicated by Δ_c. (We note LSUB2 never terminates and so there is no value for Δ_c, and this is indicated by the symbol '–'.) Details of extrapolation procedures are presented in [7]. CCM results for the spin-stiffness ρ are from [10]. Results for the magnetic susceptibility χ are determined via $\chi = \frac{dM^L}{d\lambda}$. Those results that have not, as yet, been determined at a given level of approximation have been left blank

Method	E_g/N	M	ε_e	ρ	χ	N_f^z	$N_{f_e}^z$	Δ_c
LSUB2	−0.64833	0.841	1.407	0.2310	0.0860	1	1	–
SUB2	−0.65083	0.827	1.178			∞	∞	0.799
LSUB4	−0.66366	0.765	0.852	0.2310	0.0792	7	6	0.577
LSUB6	−0.66700	0.727	0.610	0.2176	0.0765	75	91	0.763
LSUB8	−0.66817	0.705	0.473	0.2097	0.0750	1,273	2,011	0.843
LSUB10	−0.668700	0.345			0.0739	29,605	51,012	
LSUB12	−0.668978	0.339				766,220		
Extrapolated CCM	−0.66960	0.614	0.00	0.1812	0.070	∞	∞	1.03

Finally, results for the spin stiffness of the spin-half square-lattice Heisenberg model may be determined [10], and these results are also shown in Table 10.1. Again, the extrapolated value $\rho = 0.1812$ from Krüger et al. [10] compares well to the corresponding result of QMC [14] of $\rho = 0.199$.

10.3.4 Excitation Spectrum of the Spin-Half Square-Lattice XXZ-Model for the z-Aligned Model State

We now consider the excitation spectrum. We shall use the SUB2 approximation for the ground state, whereas for the excitation operator we assume

$$X = \sum_i a_i s_i^+. \tag{10.58}$$

Substitution of the expressions in Eqs. (10.48) and (10.58) for the ground- and excited-state operators, respectively, leads to the following expression [5] for the excited-state correlation coefficients

$$\frac{1}{2} n K a_k - \frac{1}{2} \sum_{\rho,r} b_r a_{k+r+\rho} = \varepsilon_e a_k. \tag{10.59}$$

This equation may also be solved by Fourier transform techniques in a similar manner presented above for the SUB2 calculation for the ground state. The result of this treatment is an expression for the excitation spectra is given by

$$\varepsilon(q) = \frac{1}{2}nK\sqrt{1 - k^2\gamma^2(q)}, \tag{10.60}$$

where b_1 is obtained from the SUB2 ket-state equations and K and k are defined by Eq. (10.54). We solve the SUB2 equations as normal at a particular value of Δ and then b_1 is substituted into Eq. (10.60) above and hence we obtain values for the excitation spectra as a function of the wave vector Eq. (10.60). Note that the excitation spectra becomes 'soft' (i.e. $\varepsilon(q) \rightarrow 0$) at the CCM SUB2 critical point at $\Delta_c \approx 0.7985$ as mentioned earlier. Results for the spectra are presented in Fig. 10.4). We see that the CCM results are in good agreement with those results of linear spin-wave theory at $\Delta = 1$. The spectra are plotted for $k_x = k_y$ and for $k_y = 0$, and we see that the CCM excitation spectrum is identical in shape to those results of SWT with a multiplicative factor of 1.1672. This agrees well with results of quantum Monte Carlo [15] that also predict a curve identical to SWT with multiplicative a factor of 1.21±0.03. Our results in thus in good agreement with SWT and QMC and this is further evidence that the CCM critical point is an indication of the quantum phase transition at $\Delta = 1$ in the 'real' system. Furthermore, the excitation spectra at this point is given by $\varepsilon(q) = \frac{1}{2}nK\sqrt{1 - \gamma^2(q)}$. This leads to a value for the spin-wave velocity v_s of $\frac{1}{2}nK$, which in turn yields a value [5] of $v_s \approx 2.335$ for the square-lattice case.

Finally, it is worth mentioning that the excitation energy may be determined directly from Eq. (10.20) in 'real space' without recourse to Fourier transform methods, although computational techniques are again necessary except for the simplest of cases. For the sake of consistency, we normally retain the same level of localised approximation for the ground and excited states. Results are presented for the XXZ-Model in Fig. 10.5 and for the Heisenberg model in Table 10.1. For the latter we see that the CCM results converge rapidly with LSUBm approximation level. Indeed, extrapolated results predict that the excitation is gapless at $\Delta = 1$,

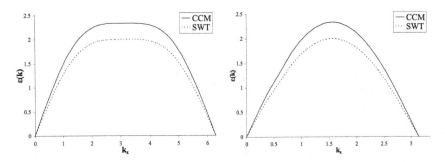

Fig. 10.4 Excitation spectra for the Heisenberg model determined at the critical point at $\Delta_c = 0.8$ for the CCM results and at $\Delta = 1$ for the spin-wave theory results. The spectra plotted on the left are for $k_x = k_y$ and those on the right are for $k_y = 0$. This agrees well with results of quantum Monte Carlo [15] that also predict a curve identical to SWT with a multiplicative factor of 1.21±0.03, respectively

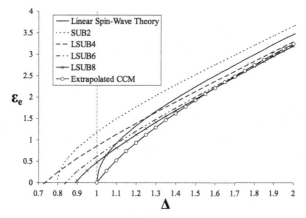

Fig. 10.5 CCM LSUB*m* results using the *z*-aligned Néel model state for the excited-state energy of the spin-half square-lattice XXZ-Model compared to linear spin-wave theory of Barnes et al. [12]

as is believed to be the case for the Heisenberg model from the results of other approximate calculations.

10.4 The Lattice Magnetisation

The lattice magnetisation is a quantity that yields an overall response of a particular quantum spin system as a whole to the externally applied magnetic field. The lattice magnetisation gives the average ordering of the spins in the direction of the externally applied magnetic field and it is defined by the equation

$$M^L \equiv -\frac{2}{N} \langle \tilde{\Psi} | \sum_i^N s_i^z | \Psi \rangle, \tag{10.61}$$

in terms of the original unrotated spin coordinates. In terms of the rotated spin coordinates, an additional factor of $(-1)^i$ is also included in Eq. (10.61). The relevant Hamiltonian for the Heisenberg model is defined by

$$\mathcal{H} = \sum_{\langle i,j \rangle} \mathbf{s}_i \cdot \mathbf{s}_j - \lambda \sum_i s_i^z , \tag{10.62}$$

where the indices i and j again run over all nearest-neighbouring lattice sites on the square lattice, although counting each bond once only, and λ indicates the strength of the external field. We must also take into account the fact that spins in the model state are explicitly allowed to cant at an angle θ to the negative and positive x-axes for the difference sublattices. This angle is treated a parameter that we treat variationally in order to obtain the best results for the energy. The total

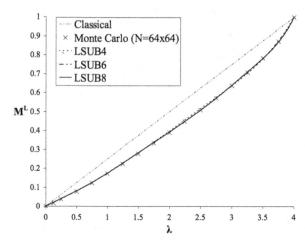

Fig. 10.6 Results for the lattice magnetisation of the spin-half square-lattice Heisenberg model in the presence of an external magnetic field of strength λ compared to results of quantum Monte Carlo [16] for $N = 64 \times 64$ and classical results

lattice magnetisation may be determined directly in a similar manner as presented above for the sublattice magnetisation. The results for the spin-half square-lattice Heisenberg model are shown in Fig. 10.6. As for the ground and excited-state energies and the sublattice magnetisations, we obtain results that converge rapidly with increasing levels of LSUBm approximation and that compare well to the results of QMC for $N = 64 \times 64$. The magnetic susceptibility may also be determined via $\chi = \frac{dM^L}{d\lambda}$ and results are shown in Table 10.1. The extrapolated CCM values of $\chi = 0.070$ for $\lambda \to 0$ again compares reasonably well to the result of QMC [17] of $\chi = 0.0669(7)$. The zero-field uniform susceptibility ($\lambda \to 0$), the ground state energy, the sublattice magnetisation, the spin stiffness, and the spin-wave velocity constitute the fundamental parameter set that determines the low-energy physics of magnetic systems. The CCM is thus able to provide a comprehensive and accurate picture of the properties of spin-half square-lattice Heisenberg model.

Appendix – Details of the Calculation of the Coefficients b_1 and $\tilde{b}_1$ in the LSUB2 Approximation

The first step in simplifying the expression for $\hat{\mathcal{H}}_2$ is to move all s^- and s^z operators to the right, using the commutation relations, noting that i and $i + \rho_0$ are on different sublattices and cannot be equal, whereas i and $i + \rho_0 + \rho_1$ are on the same sublattice, etc.

$$
\hat{\mathcal{H}}_2 = -\frac{1}{4}\sum_i \sum_{\rho_0}\left[s_i^+ s_{i+\rho_0}^+ + 4b_1^2 \sum_{\rho_1,\,\rho_3} s_{i+\rho_1}^+ s_{i+\rho_0+\rho_3}^+ s_i^z s_{i+\rho_0}^z \right.
$$

$$
+ 4b_1^2 \sum_{\rho_1} s_{i+\rho_1}^+ s_i^+ s_{i+\rho_0}^z - b_1^2 \sum_{\rho_3,\,\rho_4} s_{i+\rho_0}^+ s_{i+\rho_0+\rho_3}^+ s_{i+\rho_0+\rho_4}^+ s_i^-
$$

$$
+ 2b_1^2 \sum_{\rho_3} s_{i+\rho_0}^+ s_{i+\rho_0+\rho_3}^+ s_i^z + 2b_1^2 \sum_{\rho_4} s_{i+\rho_0}^+ s_{i+\rho_0+\rho_4}^+ s_i^z + 2b_1^2 s_i^+ s_{i+\rho_0}^+
$$

$$
- b_1^2 \sum_{\rho_1,\,\rho_2} s_i^+ s_{i+\rho_1}^+ s_{i+\rho_2}^+ s_{i+\rho_0}^- + 2\Delta b_1 \sum_{\rho_2} s_{i+\rho_0}^+ s_{i+\rho_0+\rho_2}^+ s_i^z
$$

$$
\left. + 2\Delta b_1 s_i^+ s_{i+\rho_0}^+ + 2\Delta b_1 \sum_{\rho_1} s_i^+ s_{i+\rho_1}^+ s_{i+\rho_0}^z \right]
$$

The terms with s^- on the right will give zero when acting on $|\Phi\rangle$ so after simplifying the subscripts

$$
\hat{\mathcal{H}}_2|\Phi\rangle = -\frac{1}{4}\sum_i \sum_{\rho_0}\left[s_i^+ s_{i+\rho_0}^+ + 4b_1^2 \sum_{\rho_1,\,\rho_3} s_{i+\rho_1}^+ s_{i+\rho_0+\rho_3}^+ s_i^z s_{i+\rho_0}^z \right.
$$

$$
+ 4b_1^2 \sum_{\rho_1} s_{i+\rho_1}^+ s_i^+ s_{i+\rho_0}^z + 4b_1^2 \sum_{\rho_1} s_{i+\rho_0}^+ s_{i+\rho_0+\rho_1}^+ s_i^z
$$

$$
+ 2b_1^2 s_i^+ s_{i+\rho_0}^+ + 2\Delta b_1 \sum_{\rho_2} s_{i+\rho_0}^+ s_{i+\rho_0+\rho_2}^+ s_i^z
$$

$$
\left. + 2\Delta b_1 s_i^+ s_{i+\rho_0}^+ + 2\Delta b_1 \sum_{\rho_1} s_i^+ s_{i+\rho_1}^+ s_{i+\rho_0}^z \right]|\Phi\rangle .
$$

Also $s_i^z|\Phi\rangle = -\frac{1}{2}|\Phi\rangle$ for all i so

$$
\hat{\mathcal{H}}_2|\Phi\rangle = -\frac{1}{4}\sum_i \sum_{\rho_0}\left[s_i^+ s_{i+\rho_0}^+ + b_1^2 \sum_{\rho_1,\,\rho_3} s_{i+\rho_1}^+ s_{i+\rho_0+\rho_3}^+ \right.
$$

$$
- 2b_1^2 \sum_{\rho_1} s_{i+\rho_1}^+ s_i^+ - 2b_1^2 \sum_{\rho_1} s_{i+\rho_0}^+ s_{i+\rho_0+\rho_1}^+ + 2b_1^2 s_i^+ s_{i+\rho_0}^+
$$

$$
\left. - \Delta b_1 \sum_{\rho_2} s_{i+\rho_0}^+ s_{i+\rho_0+\rho_2}^+ + 2\Delta b_1 s_i^+ s_{i+\rho_0}^+ - \Delta b_1 \sum_{\rho_1} s_i^+ s_{i+\rho_1}^+ \right]|\Phi\rangle .
$$

$$
(10.63)
$$

This is now substituted into Eq. (10.37) which is evaluated making use of the following
1.

$$
\langle\Phi|s_i^- s_j^+|\Phi\rangle = \langle\Phi|(-2s_i^z)|\Phi\rangle\delta_{ij} = \langle\Phi|\Phi\rangle\delta_{ij} = \delta_{ij},
$$

2. In the following we assume that $m \neq m'$. Typically m and m' are on different sublattices although not necessarily nearest neighbours.

$$\langle\Phi|s_l^- s_{l'}^- s_m^+ s_{m'}^+|\Phi\rangle = \langle\Phi|s_l^- s_m^+ s_{l'}^- s_{m'}^+|\Phi\rangle + \langle\Phi|s_l^-(-2s_m^z)s_{m'}^+|\Phi\rangle\delta(l'-m)$$
$$= \langle\Phi|s_l^- s_m^+ s_{m'}^+ s_{l'}^-|\Phi\rangle + \langle\Phi|s_l^- s_m^+(-2s_{m'}^z)|\Phi\rangle\delta(l'-m') + \langle\Phi|s_l^- s_{m'}^+|\Phi\rangle\delta(l'-m)$$
$$= \langle\Phi|s_l^- s_m^+|\Phi\rangle\delta(l'-m') + \langle\Phi|s_l^- s_{m'}^+|\Phi\rangle\delta(l'-m)$$
$$= \delta(l-m)\delta(l'-m') + \delta(l-m')\delta(l'-m)$$

where $\delta(i-j) \equiv \delta_{i-j,\,0} = \delta_{ij}$.
  In particular we the following two special cases of this result

$$\langle\Phi|s_l^- s_{l+\rho'}^- s_m^+ s_{m+\rho}^+|\Phi\rangle = \delta(l-m)\delta(l+\rho'-m-\rho)+\delta(l-m-\rho)\delta(l+\rho'-m)$$
$$= \delta(l-m)\delta(\rho'-\rho) + \delta(l-m-\rho)\delta(\rho+\rho')$$

and

$$\langle\Phi|s_l^- s_{l+\rho'}^- s_{m+\rho_0+\rho_3}^+ s_{m+\rho}^+|\Phi\rangle = \delta(l-m-\rho_0-\rho_3)\delta(l+\rho'-m-\rho)+$$
$$\delta(l-m-\rho)\delta(l+\rho'-m-\rho_0-\rho_3)$$
$$= \delta(l-m)\delta(\rho_0+\rho_3+\rho'-\rho)+$$
$$\delta(l-m-\rho)\delta(\rho+\rho'-\rho_0-\rho_3)$$

3.

$$\sum_{\rho_1}\delta(\rho+\rho_1) = 1$$

since there is just one ρ_1 equal to $-\rho$.

4. For the linear chain, the square lattice and the simple cubic

$$\sum_{\rho_1,\rho_2,\rho_3}\delta(\rho+\rho_1+\rho_2+\rho_3) = 3n-3$$

where n is the number of nearest neighbours, this being the number of ways that $\rho_1+\rho_2+\rho_3$ can be made equal to $-\rho$.

In (10.37) we need

$$\langle\Phi|C_l^-\hat{H}_2|\Phi\rangle = \langle\Phi|s_m^- s_{m+\rho}^-\hat{H}_2|\Phi\rangle$$

and using the above results and (10.63)

$$= -\frac{1}{4}[2+2b_1^2(3n-3)-4b_1^2n-4b_1^2n+4b_1^2$$
$$- \Delta b_1 2n + 4\Delta b_1 - \Delta b_1 2n] = 0$$
$$= -\frac{1}{2}[1-(n+1)b_1^2+2(1-n)\Delta b_1] = 0$$

which yields Eq. (10.38):

$$(n+1)b_1^2 + 2(n-1)\Delta b_1 - 1 = 0.$$

We now turn to the bra state and recall that it is not, in general, the Hermitian conjugate of the ket state. In the LSUB2 approximation, which keeps only nearest-neighbour correlations, the operator $\tilde{S}$ has the form

$$\tilde{S} = 1 + \frac{\tilde{b}_1}{2} \sum_l^N \sum_{\rho'} s_l^- s_{l+\rho'}^-, \tag{10.64}$$

where the index l runs over all sites on the lattice, ρ' runs over all nearest-neighbour sites, and $\tilde{b}_1$ is the sole bra-state correlation coefficient. $\tilde{b}_1$ can be determined using Eq. (10.14) with $C_I^+ = s_m^+ s_{m+\rho}^+$. However, it is much quicker to use a shortcut, which we show here only for the LSUB2 version.

First note that $\hat{s}_i^+ = s_i^+$ so that $\hat{C}_I^+ = C_I^+$ and Eq. (10.14) becomes

$$\langle \Phi | \tilde{S}[\hat{\mathcal{H}}, C_I^+] | \Phi \rangle = 0$$

$$\langle \Phi | \tilde{S}[\hat{\mathcal{H}}, s_m^+ s_{m+\rho}^+] | \Phi \rangle = 0$$

$$\therefore \quad b_1 \sum_m \sum_\rho \langle \Phi | \tilde{S}[\hat{\mathcal{H}}, s_m^+ s_{m+\rho}^+] | \Phi \rangle = 0$$

$$\therefore \quad \langle \Phi | \tilde{S}[\hat{\mathcal{H}}, S] | \Phi \rangle = 0 \tag{10.65}$$

since $S = b_1 \sum_m \sum_\rho s_m^+ s_{m+\rho}^+$.

Now consider

$$\frac{\partial \bar{\mathcal{H}}}{\partial b_1} = \frac{\partial}{\partial b_1} \langle \Phi | \tilde{S} e^{-S} \mathcal{H} e^S | \Phi \rangle$$

$$= \langle \Phi | \left\{ \tilde{S} \left(-e^{-S} \frac{\partial S}{\partial b_1} \right) \mathcal{H} e^S + e^{-S} \mathcal{H} \left(e^S \frac{\partial S}{\partial b_1} \right) \right\} | \Phi \rangle$$

since b_1 occurs only in S and not in $\tilde{S}$ or $\mathcal{H}$.

Also since $S = b_1 \sum_m \sum_\rho s_m^+ s_{m+\rho}^+$, it follows that $\frac{\partial S}{\partial b_1} = \frac{1}{b_1} S$ so

$$\frac{\partial \bar{\mathcal{H}}}{\partial b_1} = \frac{1}{b_1} \langle \Phi | \tilde{S}[\hat{\mathcal{H}}, S] | \Phi \rangle = 0$$

because of Eq. (10.65). Thus it is sufficient to evaluate $\bar{\mathcal{H}}$ and differentiate with respect to b_1.

We can now proceed to calculate

$$\bar{\mathcal{H}} = \langle \Phi | \tilde{S} \hat{\mathcal{H}} | \Phi \rangle$$

$$= \langle \Phi | \hat{\mathcal{H}}_0 | \Phi \rangle + \frac{\tilde{b}_1}{2} \sum_l \sum_{\rho'} \langle \Phi | s_l^- s_{l+\rho'}^- \hat{\mathcal{H}}_2 | \Phi \rangle$$

The first term is E_g evaluated earlier:

$$\langle \Phi | \hat{\mathcal{H}}_0 | \Phi \rangle = -\frac{Nn}{8}(\Delta + 2b_1)$$

while the term inside the summation in the second term was evaluated in determining b_1 and is given by (using m and ρ instead of l and ρ')

$$\langle \Phi | s_m^- s_{m+\rho}^- \hat{\mathcal{H}}_2 | \Phi \rangle = -\frac{1}{2}[1 - (n+1)b_1^2 - 2(n-1)\Delta b_1].$$

Thus, carrying out the summations,

$$\bar{\mathcal{H}} = -\frac{Nn}{8}(\Delta + 2b_1) - \frac{\tilde{b}_1 Nn}{4}[1 - (n+1)b_1^2 - 2(n-1)\Delta b_1].$$

Differentiating with respect to b_1 and setting this equal to 0 yields Eq. (10.43).

$$-1 + \tilde{b}_1[(n+1)2b_1 + 2(n-1)\Delta] = 0.$$

References

1. Bishop, R.F.: Theor. Chim. Acta **80**, 95 (1991)
2. Roger, M., Hetherington, J.H.: Phys. Rev. B **41**, 200 (1990)
3. Roger, M., Hetherington, J.H.: Europhys. Lett. **11**, 255 (1990)
4. Bishop, R.F., Parkinson, J.B., Xian, Y.: Phys. Rev. B **44**, 9425 (1991)
5. Bishop, R.F., Parkinson, J.B., Xian Y.: Phys. Rev. B **46**, 880 (1992)
6. Zeng, C., Farnell, D.J.J., Bishop, R.F.: J. Stat. Phys. **90**, 327 (1998)
7. Bishop, R.F., Farnell, D.J.J., Krüger, S.E., Parkinson, J.B., Richter, J., Zeng, C.: J. Phys. Condens. Matter **12**, 7601 (2000)
8. Farnell, D.J.J., Gernoth, K.A., Bishop, R.F.: J. Stat. Phys. **108**, 401 (2002)
9. Farnell, D.J.J., Bishop, R.F.: arxi.org/abs/cond-mat/0311126 (1999)
10. Krüger, S.E., Darradi, R., Richter, J., Farnell, D.J.J.: Phys. Rev. B **73**, 094404 (2006)
11. Rosenfeld, J., Ligterink, N.E., Bishop, R.F.: Phys. Rev. B **60**, 4030 (1999)
12. Barnes, T., Kotchan, D., Swanson, E.S.: Phys. Rev. B **39**, 4357 (1989)
13. Sandvik, A.W.: Phys. Rev. B **56**, 11678 (1997)
14. Makivic, M.S., Ding, H.Q.: Phys. Rev. B **43**, 3562 (1991)
15. Chen, G., Ding, H.-Q., Goddard, W.A.: Phys. Rev. B **46** 2933 (1992)
16. Richter, J., Schulenburg, J., Honecker A.: Quantum magnetism in two dimensions: From semi-classical Néel order to magnetic disorder. In: Schollwöck, U., Richter, J., Farnell, D.J.J., Bishop, R.F. (eds.) Quantum Magnetism. Lecture Notes in Physics, vol. 645, pp. 84–153. Springer, Berlin (2004)
17. Singh, R.R.P., Gelfand, M.P.: Phys. Rev. B **52**, R15695 (1995)

Chapter 11
Quantum Magnetism

Abstract Many quantum spin systems of spin quantum number $s > 1/2$ or in spatial dimensions greater than 1D cannot be solved exactly. One source of this "lack of integrability" comes from the competition between different bonds, as in quantum frustration. We begin this final chapter by considering the application of approximate methods to one-dimensional models such as the spin-half $J_1 - J_2$ model, and the spin-one Heisenberg and biquadratic models. The properties of the spin-half Heisenberg model for Archimedean lattices such as the square (unfrustrated) lattice, and the triangular and Kagomé (frustrated) lattices are listed for a variety of approximate techniques. The phenomenon of "order-from-disorder" is investigated in the context of the triangular lattice antiferromagnet in the presence of an external field. Finally, the properties of the square-lattice $J_1 - J_2$ model and the Shastry-Sutherland model are studied. The chapter shows how the application of a range of approximate techniques, in addition to the few isolated exact results, can provide a comprehensive and compelling description of the ground-state properties of a wide range of quantum spin systems.

11.1 Introduction

In previous chapters we saw that the spin-half one-dimensional Heisenberg model may be solved exactly by using the Bethe Ansatz and that the Néel order of the classical ground state is removed by quantum fluctuations. However, there are other ways in which the classical ordering may be replaced by states of quantum order that have no classical counterpart or by quantum disorder.

One possible situation in which this can happen is in frustrated systems [1]. The term 'frustration' indicates that different bonds in the Hamiltonian compete against each other. A simple example is a triangle of spins with antiferromagnetic exchange interactions. The spins wish to align antiparallel to their neighbours but this is not possible for all bonds (connected pairs of atoms) simultaneously. Classically, this means that the energies for these different bonds are higher than for their corresponding unfrustrated counterparts.

Bipartite lattices (such as the square lattice) with nearest neighbour antiferromagnetic exchange interactions are unfrustrated since in the classical Néel ground state

Parkinson, J.B., Farnell, D.J.J.: *Quantum Magnetism*. Lect. Notes Phys. **816**, 135–152 (2010)
DOI 10.1007/978-3-642-13290-2_11   © Springer-Verlag Berlin Heidelberg 2010

each spin is aligned antiparallel to all its neighbours, and the energy associated with each bond is $-JS^2$. (Bipartite lattices are those lattices that may be divided into two interpenetrating sublattices in which nearest neighbours to sites on one sublattice are always on the other sublattice and vice versa.) By contrast, the classical ground state of the spin-half Heisenberg model on the frustrated triangular lattice has spins at an angle of 120° to each other. The classical energy of a single antiferromagnetic bond in the Hamiltonian is now $-\frac{1}{2}JS^2$, i.e. exactly half of its unfrustrated counterpart on a bipartite lattice.

The field of quantum magnetism has been strongly influenced recently by the discovery of many new low-dimensional magnetic materials. These new materials go from zero dimensions, such as magnetic clusters and molecules, to magnetic materials with underlying three-dimensional crystallographic lattices. In this final chapter, we discuss how the application of the approximate modern-day techniques of quantum many-body theory may be used predict and understand these systems, especially where they exhibit either 'novel' ground states that have classical counterpart or quantum disorder. The interested reader is referred to [2] for a review of the field of quantum magnetism.

11.2 One-Dimensional Models

The subject of one-dimensional magnetism is interesting because the elementary excitations are constrained to interact strongly because of the low-dimensional nature of lattice. Furthermore, as we have seen, one-dimensional systems are sometimes amenable to exact solution by using methods such as the Bethe Ansatz and the Jordan-Wigner transformation. Indeed, much of this book has been concerned with one-dimensional systems and how they may be treated using a variety of exact and approximate techniques. We shall describe three further one-dimensional models of particular interest that are not generally amenable to exact solution at all points. These are (1) the spin-half chain with nearest-neighbour exchange J_1 and next nearest exchange J_2, which is frustrated for $J_2 > 0$, (2) the spin-one chain with nearest-neighbour antiferromagnetic Heisenberg exchange, and (3) the spin-one chain with nearest-neighbour exchange J_1 and also a nearest-neighbour biquadratic exchange J_2. These systems demonstrate fascinating physics that often has no classical counterpart. There are many other one-dimensional systems that also have interesting physics and are realisable experimentally, such as spin 'ladder' systems and other 'quasi-1D' systems, and other models with more than two-body spin exchange. We shall not describe these here nor attempt to give a complete list and the interested reader is referred to Refs. [2–5] for reviews of one-dimensional quantum spin systems.

11.2.1 The Spin-Half J_1–J_2 Model on the Linear Chain

We now consider a spin-half model on the one-dimensional linear chain with periodic boundary conditions with both nearest- and next-nearest-neighbour bonds of strength J_1 and J_2, respectively. This system is known as the spin-half J_1–J_2 model [6–19]. Much interest in this model was rekindled by the discovery of the spin-Peierls material CuGeO$_3$, although it is relevant to many other magnetic materials also. The Hamiltonian for this model is given by

$$H = \frac{J_1}{2} \sum_{i,\rho_1} \mathbf{s}_i \cdot \mathbf{s}_{i+\rho_1} + \frac{J_2}{2} \sum_{i,\rho_2} \mathbf{s}_i \cdot \mathbf{s}_{i+\rho_2}, \qquad (11.1)$$

where the index i runs over sites on the lattice counting, ρ_1 runs over all nearest-neighbours to i, and ρ_2 runs over all next-nearest-neighbours to i. It is also convenient to write the bonds strengths as $J_1 = J \cos(\omega)$ and $J_2 = J \sin(\omega)$, and then choose $J = 1$ for simplicity. The ground-state properties of this system have been studied using methods such as exact diagonalizations [8, 15], the density matrix renormalisation group (DMRG) [9–13, 18], CCM [16–19], and field-theoretical approaches [13] (see Refs. [13, 14] for a general review). DMRG results [9–13, 18] provide the most accurate results for this model. Indeed, the DMRG generally provides (in the absence of exact results) the benchmark for one-dimensional systems to which other methods are compared. QMC results can also be extremely accurate in those cases in which it is applicable, in particular in the absence of frustration.

For the J_1–J_2 model in the region $J_2/J_1 > 0$, the nearest-neighbour (J_1) and next-nearest-neighbour interactions (J_2) compete and so they are 'frustrated.' At the Majumdar-Ghosh point $J_2/J_1 = \frac{1}{2}$ it is straightforward to prove [6, 7] that there are two degenerate exact dimer-singlet product ground states, one of which (with appropriate normalisation) is given by

$$|\Psi\rangle = \{|\downarrow_1\uparrow_2\rangle - |\uparrow_1\downarrow_2\rangle\} \otimes \{|\downarrow_3\uparrow_4\rangle - |\uparrow_3\downarrow_4\rangle\} \otimes \cdots$$
$$\otimes\{|\downarrow_{N-1}\uparrow_N\rangle - |\uparrow_{N-1}\downarrow_N\rangle\}. \qquad (11.2)$$

Another valid ground state is that of Eq. (11.2) with all sites translated by one unit. We note that each two-site 'dimer' state $\{|\downarrow_i\uparrow_{i+1}\rangle - |\uparrow_i\downarrow_{i+1}\rangle\}$ is simply the ground state of the $N = 2$ antiferromagnetic Heisenberg system presented in Chap. 2. The effect of frustration is to break the symmetry of the lattice via the spontaneous creation of this product state of dimers for the extended system, namely, for $N \to \infty$. This is a purely quantum result that has no classical counterpart and so is noteworthy. An energy gap to the first excited state exists and is of magnitude $0.234J_1$ (from [4] p. 107).

The phase diagram of this system is shown in Fig. 11.1. The system is ferromagnetically ordered in the region $-\pi \leq \omega \leq -\frac{\pi}{2}$. Furthermore, we have the unfrustrated Heisenberg antiferromagnet, where the exact solution is provided by the Bethe Ansatz at $J_2/J_1 = 0$ (i.e., $\omega = 0$). Indeed, this (gapless) phase extends over the entire region $-\frac{\pi}{2} \leq \omega \leq 0$ in which next-nearest-neighbour bonds do not compete with their nearest-neighbour counterparts. We remark that the system is still gapless up until the point $J_2/J_1 = 0.2411(1)$, at which point the model exhibits a transition to the two-fold degenerate gapped dimerized phase with an

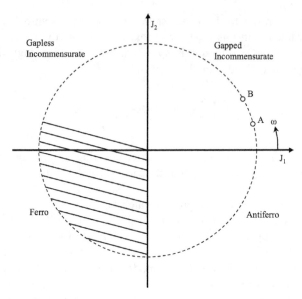

Fig. 11.1 The phase diagram of the spin-half J_1–J_2 model on the linear chain. There is a transition to a two-fold degenerate gapped ground state at *Point A*. The dimerised state is the exact ground state at the Majumdar-Ghosh *Point B*

exponential decay of the correlation function $\langle s_i \cdot s_j \rangle$ [8–11, 13, 14]. There is a Lifshitz point for $J_2/J_1 \approx 0.5206$ at which point incommensurate spiral correlations occur. From DMRG studies [13], the excitation energy gap persists well into the region $J_2/J_1 > 1/2$. Finally, a gapless spiral phase probably occurs in the region $\frac{\pi}{2} \leq \omega \lesssim 2.9$. At a point very close to $J_2/J_1 = -0.25$ (i.e., $\omega \approx 2.9$), there is a transition to the ferromagnetic ground state.

11.2.2 The s = 1 Heisenberg Model on the Linear Chain

By contrast with the $s = 1/2$ Heisenberg model on the linear chain, no exact solution for the equivalent spin-one one-dimensional Heisenberg model has as yet been found. However, extremely accurate DMRG calculations have been performed [20] for this model giving a value for the ground-state energy of $E_g/N = -1.401484038971(4)$. Exact diagonalisations using the SpinPack Code of Joerg Schulenburg [21] (extrapolated using the rule: $E_g = a + bN^{-2}$) yield a value for the ground-state energy that agrees with DMRG to six decimal places, whereas, for example, the extrapolated CCM results [22] were found to yield a value of $E_g/N = -1.403737$. We remark again that DMRG generally provides the most accurate results in 1D. In 1983, Haldane [23, 24] postulated that this system might contain a gap, and exact diagonalisations were the first independent calculations to confirm the presence of this gap. However, it was DMRG calculations that conclusively showed that there is an excitation energy gap of magnitude 0.41050(2) in this system. (Modern-day exact diagonalisations again using the SpinPack code

[21] now yield a value that agrees with this value to five decimal places.) Hence, the spin-one anisotropic Heisenberg model on the linear chain is in the 'Haldane phase,' in which the amount of long-range Néel-ordering is zero and the excitation spectrum is gapped. We remark that this is in stark contrast to the spin-half model which is gapless for the quantum system and also to its classical counterpart which is Néel-ordered. Furthermore, we note that conventional spin-wave theory predicts that the excitation energy of the Heisenberg model on the infinite linear chain for both the spin-half *and* spin-one cases is gapless. This is therefore a model for which the application of SWT of the sort presented in this book fails. This result for the disordered state of the $s = 1$ antiferromagnet reflects a more general property of integer-s chains that demonstrate only short-range, exponentially decaying AF correlations. For the spin-one XXZ-Model, the phase transition from the Néel-like phase occurs for a value of anisotropy $\Delta = 1.167 \pm 0.007$ [25], which is again in contrast to the spin-half system for which a transition to a gapless co-planar regime occurs at exactly $\Delta = 1$. We refer the interested reader to [2] (Chap. 1) for further information regarding the spin-one Heisenberg model on the linear chain.

11.2.3 The s = 1 Heisenberg-Biquadratic Model on the Linear Chain

We now consider a chain of spins with $s = 1$ and nearest neighbour interactions of a more complicated type. The Hamiltonian has the form

$$ H = \frac{1}{2} \sum_{i,\rho_1} [J_1 \, \mathbf{s}_i \cdot \mathbf{s}_{i+\rho_1} + J_2 \, (\mathbf{s}_i \cdot \mathbf{s}_{i+\rho_1})^2]. \tag{11.3} $$

Again we write the bonds strengths as $J_1 = J \cos(\omega)$ and $J_2 = J \sin(\omega)$, and then choose $J = 1$.

Clearly the model in the previous section is an important particular case of this Hamiltonian in which $\omega = 0$. The first term is the normal isotropic Heisenberg exchange interaction and has a *bilinear* form. The second term has a *biquadratic* form and so the model is known as the Bilinear-Biquadratic model, or Heisenberg-Biquadratic model.

The phase diagram can be presented in a form similar to that given earlier and is shown in Fig. 11.2. The system is integrable using the Bethe Ansatz at the points marked T, at $\omega = -\pi/4$ [26, 27], and S, at $\omega = \pi/4$ [28, 29]. The exact ground state energy is also known at the point marked B, at $\omega = -\pi/2$ [30–34], where there is an exact mapping of some states onto a spin-half chain with anisotropic exchange. The exact ground state is also known at A, $\omega = \tan^{-1}(1/3)$ [35, 36] and has a dimerised form similar to the Majumdar-Ghosh state mentioned earlier.

The point H, at $\omega = 0$ is the $s = 1$ linear chain Heisenberg model discussed in the previous section, which was predicted by Haldane, on the basis of a field-theoretical calculation, to have a gap. Numerical work has subsequently strongly supported this prediction.

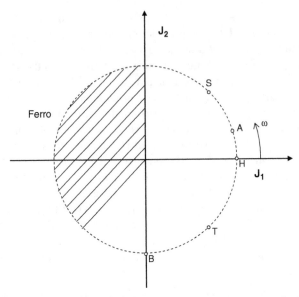

Fig. 11.2 The zero-temperature phase diagram of the spin-1 Bilinear-Biquadratic model on the linear chain. The various marked points, referred to in the text, correspond to values of ω at which exact results are known

The nature of the states for other values of ω, for which exact results are not known, has been discussed by Chubukov [37, 38] and by Fáth and Sólyom [39, 40] and early numerical work was carried out by Xiang and Gehring [41]. The system is believed to be gapless for $\pi/4 \geq \omega \geq \pi/2$ and to have a gap for most of the remaining non-ferromagnetic regime, although the form of the states is different for $-3\pi/4 > \omega > -\pi/4$ and $-\pi/4 > \omega > \pi/4$.

11.3 The $s = 1/2$ Heisenberg Model for Archimedean Lattices

Two-dimensional magnetism is a fascinating subject because the physics is driven by the often complex nature of the underlying crystallographic lattice, the number and range of bonds in this lattice, and spin quantum numbers of atoms localised to the sites on the lattice. There are very few exact results for quantum spin systems on two-dimensional lattices so the application of approximate methods is crucial to their understanding.

These systems display a wide variety of behaviour from semi-classical Néel ordering, two-dimensional quantum 'spirals' to valence-bond crystals/solids and spin liquids. Exact diagonalisations of systems with finite N are a very useful tool, and often expectation values may be reliably extrapolated in the limit of $N \to \infty$. However, consideration of the excited states as a function of total s also gives important information regarding the ordering of the ground state because the lowest lying energies in each sector of total s should scale as $s(s + 1)$ for Néel-ordered systems.

The interested reader is referred to Ref. [2] (Chap. 2) and [1] (Chap. 5) for more details of two-dimensional quantum magnetism.

Key also to understanding two-dimensional problems is the concept of the unit cell and the Bravais lattice. The unit cell contains a number of sites at specific positions (given by the 'primitive' lattice vectors) that are replicated at all possible multiples of the Bravais lattice vectors. Thus, for example, we have a single site in the unit cell for the linear chain, say, at position $(0, 0)$ and a single Bravais lattice vector $\hat{a} = (1, 0)^T$. The lattice is formed by translating the single site in the unit cell by all integer multiples of $\hat{a}$. Two-dimensional lattices have two Bravais lattice vectors. For example the square lattice has a single site in the unit cell and the lattice vectors are $\hat{a} = (1, 0)^T$ and $\hat{b} = (0, 1)^T$. The triangular lattice is given by vectors $\hat{a} = (1, 0)^T$ and $\hat{b} = (1/2, \sqrt{3}/2)^T$ and so on for other lattices.

Of special interest in two spatial dimensions are the Archimedean lattices in which the distance between all nearest-neighbour lattice sites is equal to one. The Archimedean lattices may be divided into two broad classes, namely, those that are geometrically frustrated and those that are not. Figure 11.4 illustrates a number of common Archimedean lattices. Lattices (a)–(c) in Fig. 11.3 are the square, honeycomb, and CAVO lattices, and these lattices are all bipartite because we may

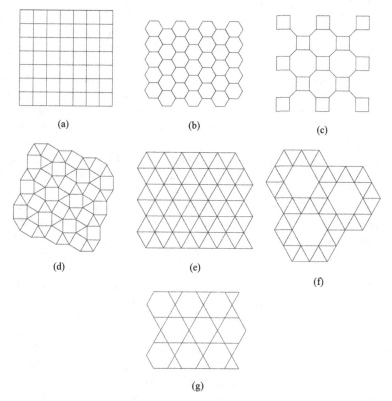

Fig. 11.3 Archimedean Lattices: (**a**) Square; (**b**) Honeycomb; (**c**) CAVO; (**d**) SrCu(BO$_3$)$_2$; (**e**) Triangle; (**f**) Maple Leaf; and (**g**) Kagomé

divide their sites into two sublattices such that all nearest neighbours of a site on one sublattice lie on the other sublattice. Bipartite lattices are not frustrated. The CAVO lattice shown in Fig. 11.3c is the underlying lattice for the magnetic material CaV_4O_9.

Results for the ground-state energy of these lattices are shown in Table 11.1. Since they are not frustrated quantum Monte Carlo (QMC) techniques can be used without difficulty. In general, in those cases in which it may be applied, QMC provides the most accurate results and so they are the benchmark to which other methods are compared. However, good agreement is observed for the ground-state energy between all approximate methods for the results presented in Table 11.1 for the bipartite lattices (a)–(c). The lowest lying energies in each sector of total spin s also scale as $s(s + 1)$, thus indicating that each of these lattices has a Néel-like ordering. The order parameter (with a factor of 1/2 compared to results in Chap. 11) is also shown in Table 11.1 and the results indicate that the amount of Néel ordering is reduced compared to the classical case. Again, good correspondence between the different approximate methods is observed. Finally, we remark that the excitation energy gap is believed to be zero for all of the spin-half bipartite Archimedean lattices. A review of the properties of the spin-half square-lattice antiferromagnet, for example, is given in [42].

By contrast, lattices (d)–(g) all exhibit frustration. Results for the ground-state energies are shown in Table 11.2. In these cases, no exact sign rule has yet been found and so the application of QMC is severely limited. However, once again good correspondence between the results of the different approximate methods is seen. For lattices (d)–(f), namely, SrCuBO (the underlying lattice of $SrCu(B0)_3$), the triangular and maple-leaf lattices, it is found that the ground state has Néel-like ordering, indicated by the fact that the lowest lying energies with total spin s scale as $s(s + 1)$, as indicated above (see [2], Chap. 2). Note that classically nearest-neighbours align at angles of 120° to each for the Heisenberg antiferromagnet on the triangular lattice. There are two sublattices for the SrCuBO lattice, three sublattices for the triangular lattice, and six sublattices for the maple-leaf lattice. Again, results of approximate methods for the sublattice magnetisation shown in Table 11.2 indicate that these systems are semi-classically ordered, albeit by an reduced amount compared to the classical system due to quantum fluctuations. A review of the triangular lattice antiferromagnets and relevant magnetic materials is given in [62]. The SrCuBO lattice (d) shown in Fig. 11.3 is a special case of the Shastry-Sutherland model discussed below.

The Kagomé-lattice antiferromagnet of lattice (g) in Fig. 11.3 is also interesting because classically it has an infinite number of ground states. Furthermore, consideration of the lowest lying energies in each sector of total s show that these do not scale as $s(s + 1)$ thus indicating this it is not Néel-ordered. This is also supported by results for the sublattice magnetisation, which is believed to be zero in the infinite lattice limit ($N \rightarrow \infty$). The Kagomé-lattice antiferromagnet is related to the broader topic of 'spin ice' materials that exhibit non-zero entropies even at zero temperature [1]. The energy gap to the first excited state is again believed to be zero for lattices (d)–(g).

Table 11.1 Approximate results for ground-state energies E_g per bond and the sublattice magnetisation $M(= \langle s^z \rangle)$ of the (unfrustrated) spin-half Heisenberg antiferromagnet for the square (a), honeycomb (b) and CAVO (c) lattices

Lattice		QMC	SWT	CSE	ED	CCM
Square (a)	$\frac{E_g}{\text{bond}}$	-0.334719(3) [43]	-0.334995 [44, 45]	-0.33465(5) [46]	-0.3350	-0.3348
	M	0.3070(3) [43]	0.3069 [44, 45]	0.307(1) [46]	0.3173	0.307
Honey (b)	$\frac{E_g}{\text{bond}}$	-0.3630 [47]	-0.365929 [48]	-0.3629 [49]	-0.3632	-0.363155 [50]
	M	0.235 [47]	0.2418 [48]	0.266 [49]	0.2788	0.274066 [50]
CAVO (c)	$\frac{E_g}{\text{bond}}$	-0.3630 [51]	-0.3588 [52]	-0.3629 [53]	-0.3682	-0.36888 [54]
	M	0.178(8) [51]	0.212 [52]	–	0.2303	0.204 [54]

Numbers in brackets indicate the error.

QMC=quantum Monte Carlo; SWT=spin-wave theory; CSE=cumulant series expansion; ED=exact diagonalisations; and CCM=coupled cluster method. All ED results all taken from [2], Chap. 2. CCM results for the square lattice taken from Chap. 11 (with a factor of 1/2 for M).

Table 11.2 Approximate results for ground-state energies per bond and sublattice magnetisations $M(= \langle s^z \rangle)$ of the (frustrated) spin-half Heisenberg antiferromagnet for the SrCuBO (d), triangular (e), maple-leaf (f), and Kagomé (g) lattices

Lattice		SWT/SBMF	CSE	ED	CCM
SrCu(BO$_3$)$_2$ (d)	$\frac{E_g}{bond}$	−0.231 [55]	−0.231 [56]	−0.2310	−0.2311
	M	0.203 [55]	0.200 [56]	0.2280	0.211
Triangle (e)	$\frac{E_g}{bond}$	−0.1823 [57]	−0.1842 [58]	−0.1842	−0.184147 [50]
	M	0.2387 [57]	0.20 [58]	0.193	0.189331 [50]
Maple Leaf (f)	$\frac{E_g}{bond}$	−0.20486 [59]	–	−0.2171	−0.2115
	M	0.154 [59]	–	0.0860	0.063
Kagomé (g)	$\frac{E_g}{bond}$	−0.2353 [60]	–	−0.2172	−0.2126 [61]
	M	–	–	∼ 0	Consistent with 0 [61]

SWT/SBMF=spin-wave theory/Schwinger-Boson mean–field; CSE=cumulant series expansion; ED=exact diagonalisations; and CCM=coupled cluster method.

All ED results all taken from [2], Chap. 2. CCM results for the maple-leaf lattice calculated by the authors.

Thus we see that the application of a range of techniques can help greatly in understanding the basic properties of a range of frustrated and unfrustrated quantum spin systems in two spatial dimensions.

11.4 Spin Plateaux

Another interesting topic in the field of quantum magnetism is that of 'spin plateaux' in which the total lattice magnetisation remains constant over a range of values of of an externally applied magnetic field. For finite-sized systems, changes from one plateau to the next may be due to transitions between different quantum states as the external magnetic field is increased. However, it is found that that spin plateau can also occur in certain cases in the infinite-lattice limit ($N \rightarrow \infty$).

An example of this is given by the spin-half (Archimedean) triangular-lattice Heisenberg antiferromagnet in the presence of an external magnetic field. The relevant Hamiltonian is

$$H = \frac{J}{2} \sum_{i,\rho} \mathbf{s}_i \cdot \mathbf{s}_{i+\rho} - \lambda \sum_i s_i^z, \qquad (11.4)$$

where the index i runs over all lattice sites on the triangular lattice and ρ runs over all nearest-neighbours to i. J is the exchange constant which we take to be $+1$ for simplicity and $\lambda = g\beta H$ where H is the strength of the applied external magnetic field. Again for simplicity we work with units in which $g\beta$ is unity so that λ is equivalent to H.

As we have seen above, the quantum ground state at $\lambda = 0$ is semi-classically ordered, although the order is reduced by quantum fluctuations from the classical value. The classical behaviour is that the nearest-neighbour spins align at angles of $120°$ to each other for this Heisenberg antiferromagnet on the (tripartite) triangular lattice at $\lambda = 0$. In the presence of an externally applied magnetic field ($\lambda > 0$), the classical picture indicates that the spins will cant at various angles and that at a 'saturation' value of $\lambda = \lambda_s$ ($\equiv 4.5$) all spins align with the field. The magnetisation saturates to a maximum value $M = M_s = Ns$, where s is the spin magnitude, at this point.

In contrast to the behaviour of the square-lattice antiferromagnet in an external magnetic field discussed in Chap. 10, the quantum behaviour of the spin-half triangular-lattice antiferromagnet [63–73] is very different to that of the classical model. In particular, a magnetisation plateau is observed at $M/M_s = \frac{1}{3}$ over a significant region of λ, as shown in Fig. 11.4. The range of this plateau has been estimated by spin-wave theory [68, 69] to be given by $1.248 < \lambda < 2.145$, by exact diagonalisations [63, 66, 67] to be given by $1.38 < \lambda < 2.16$ and by the CCM [74] to be $1.37 < \lambda < 2.15$ and these results are shown in the figure. We note again that application of the QMC method (which gives very accurate results for bipartite lattices) to the case of the triangular is severely limited by the 'sign problem' due to frustration.

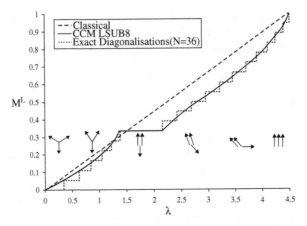

Fig. 11.4 Results for the total lattice magnetisation M as a fraction of the maximum magnetisation M_s of the spin-half triangular-lattice Heisenberg antiferromagnet in the presence of an external magnetic field of strength λ. CCM results are compared to those results of exact diagonalisations obtained using Joerg Schulenburg's SpinPack code. The arrows illustrate the actual spin directions

Note that the plateau state (uud) is collinear (i.e. all spins are aligned parallel or antiparallel to the direction of the magnetic field). This plateau is an illustration of the phenomenon of 'order from disorder' in which quantum fluctuations tend to favour collinear states. Furthermore, it is a purely quantum effect because no such plateau exists for the classical system.

Finally, recent experimental evidence [75] for the magnetic material Cs_2CuBr_4 suggests that a series of plateaux might exist at values of M/M_s equal to 1/3, 1/2, 5/9 and 2/3. This might be due to unit cells of differing size for the different plateaux. At present theory has only predicted the first of these at 1/3. However, the topic of these possible higher plateau is beyond the scope of this book, and the interested reader is referred to [2] (Chap. 2) for more information regarding spin plateaux generally.

11.5 The Spin-Half J_1–J_2 Model on the Square Lattice

We have seen already that the addition of frustrating next-nearest-neighbour bonds to the Heisenberg antiferromagnet with purely nearest-neighbour interactions can lead to a dimerised ground state for the spin-half linear-chain J_1–J_2 model. However, we may also consider what the effects of frustrating next-nearest-neighbour terms are for the equivalent square-lattice model and we can see if they again lead to interesting behaviour. Recent interest in this model comes also from studies of various layered magnetic materials, such as Li_2VOSiO_4, Li_2VOGeO_4, $VOMoO_4$, and $BaCdVO(PO_4)_2$.

The relevant Hamiltonian for this model is again given by Eq. (11.1), although i runs over all sites on the square lattice in this case. ρ_1 again runs over all nearest-neighbours to i (along the sides of the squares) and ρ_2 runs over all next-nearest-neighbours (along the diagonals of the squares), and we again write the bond strengths as $J_1 = \cos(\omega)$ and $J_2 = \sin(\omega)$. Amongst the most useful appoximate methods that have been used to simulate the properties of this systems are the CCM [76–80], series expansions [81–85], exact diagonalisations [86–90], and hierarchical mean-field calculations [91]. However, other approximate methods have also been used and the interested reader is referred to [90] for a recent review.

It is found that the system is ferromagnetically ordered in the region $-\pi \leq \omega \leq -\frac{\pi}{2}$. At $\omega = 0$ ($J_2 = 0$) we have the unfrustrated Heisenberg antiferromagnet which is Néel ordered as discussed above. This Néel-ordered phase extends over the entire region $-\frac{\pi}{2} \leq \omega \leq 0$ in which next-nearest-neighbour bonds do not compete with their nearest-neighbour counterparts. The system remains Néel ordered up until the point $J_2/J_1 \approx 0.4$. Unlike the one-dimensional J_1–J_2 system, there is no equivalent spiral phase in 2D for large J_2/J_1.

For $J_2/J_1 \gtrsim 0.6$, CCM, mean-field, series expansion and exact diagonalisation (ED) results all indicate that the system is Néel ordered over next-nearest neighbours and this ordering is one in which spins form stripes (either on the columns or rows). Furthermore, there is an intermediate regime for $0.6 \gtrsim J_2/J_1 \gtrsim 0.4$ between the two types of nearest- and next-nearest-neighbour Néel ordering. Series expansion and ED suggest that there is a spin gap in this region. The nature of the ground state in this intermediate regime has been postulated to be a valence-bond crystal formed of dimers or 4-site 'plaquettes' [81, 83, 91] or a resonating valence bond state [92, 93]. Results for the divergence or enhancement of the generalized susceptibilities obtained by CCM and ED [80] approaching the intermediate regime from the Néel phase are particularly suggestive of ground states that break translational symmetry. These calculations do not suggest a spatially homogeneous spin-liquid state without any long-range order. However, it is fair to say that the behaviour of the spin-half J_1–J_2 model on the square lattice is still not completely well-understood in this intermediate regime and that a general consensus as to the nature of the ground state has yet to emerge. At or very close to $J_2/J_1 = -0.5$ (i.e., $\omega \approx 2.68$), there is a transition from the striped state to the ferromagnetic ground state (possibly with a brief intermediate phase).

11.6 The Shastry-Sutherland Antiferromagnet

There is another model for an underlying square lattice also with frustrating next-nearest-neighbour bonds that demonstrates dimer order more conclusively and it is called the Shastry-Sutherland antiferromagnet [94]. The Shastry-Sutherland antiferromagnet is a spin-half Heisenberg model on an underlying square lattice with antiferromagnetic nearest-neighbour bonds J_1 and with one antiferromagnetic next-nearest-neighbour diagonal bond J_2 in every second square (plaquette), as shown in Fig. 11.5. Interest in this model has been renewed by the discovery of the magnetic material SrCu(BO$_3$)$_2$ [55] that can be understood in terms of the Shastry-Sutherland model. The ground state of this model in the limit of small frustration $J_2/J_1 \ll 1$ and large frustration $J_2/J_1 \gg 1$ is well understood.

At the point $J_2/J_1 = 1$ it is equivalent to the Archimedean lattice (d) denoted SrCu(BO$_3$)$_2$ in Fig. 11.3, and Table 11.2 indicates that there is strong evidence

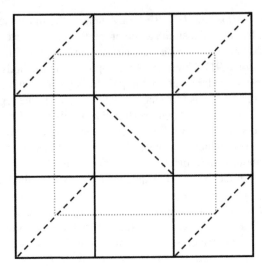

Fig. 11.5 The nearest-neighbour bonds (*solid lines*) of strength J_1 and the next-nearest-neighbour diagonal bonds (*dashed lines*) of strength J_2 for the Shastry-Sutherland model. The geometric unit cell is shown by the *square with the grey dotted lines*

that it is Néel-ordered at this point. This model has been treated previously by Schwinger boson mean-field theory [55], exact diagonalization of small lattices [95, 96], series expansions [56, 97–99], the renormalization group [99], a gauge-theoretical approach [100], and the CCM [102, 103]. A recent review can be found in [101]. We know from these studies that the physics of the quantum model is similar to that of its classical counterpart for small $J_2 < J_1$, i.e., we have semi-classical Néel long-range order.

Furthermore, for large J_2, just as in the case of the one-dimensional J_1–J_2 model, this model has a simple exact dimer-singlet product ground state [94] given by dimer singlets on the diagonal bonds indicated by the dashed lines in Fig. 11.5. However, for the Shastry-Sutherland model it is remarkable that this simple dimer product ground state is the exact ground state over a wide range of $J_2/J_1 (> J_2/J_1|_c)$, where $J_2/J_1|_c$ is believed to be $\approx (1.465 \pm 0.025)$. This is associated with the fact that, in contrast to the one-dimensional J_1–J_2 model, the dimer-singlet product ground state does not break the translational symmetry of the underlying lattice. The ground-state energy per site of this dimer-singlet product state state is $E_{\mathrm{dimer}}/N = -3J_2/8$.

The ground-state phase at intermediate values, $1 \lesssim J_2/J_1 \lesssim 2$, is still a matter of discussion. However, the classical ground state in the limit of intermediate to large values of J_2/J_1 is a two-dimensional 'spiral' and this state can be used a starting point for the application of the CCM to this model [102, 103]. CCM results for this model suggest strongly that no intermediate regime exists between the Néel regime for $J_2/J_1 < J_2/J_1|_c$ and that of the exact dimer product state. Hence, they suggest that this transition is direct and furthermore that it is most likely of first order.

11.7 Conclusions

In this book, we have presented methods of treating quantum spin systems via exact and approximate methods. We have seen that exact solutions are usually only available for unfrustrated spin-half systems with nearest-neighbour interactions in one-dimension. In principle exact solutions provide all of the information that is needed about the system in question, although it may be difficult to extract this information in practice.

However, various isolated exact solutions also occur for other systems, such as the Majumdar-Ghosh point for the (frustrated) spin-half one-dimensional J_1–J_2 model at $J_2/J_1 = 0.5$, the spin-one biquadratic model, and the two-dimensional Shastry-Sutherland model with large next-nearest-neighbour exchange interactions. These are exceptions, however, and exact solutions are rare for systems of quantum spin number $s \geq 1/2$, lattices of higher spatial dimensionality, and frustrated spin systems. Consequently approximate methods are necessary in order to understand the properties and behaviour of the majority of spin systems for which exact results do not exist.

For one-dimensional and quasi-one-dimensional systems, the method of choice is the DMRG method, which provides results of high accuracy. For higher dimensions the QMC method provides excellent results in those cases for which it can be applied, namely, in the absence of frustration. Exact diagonalisation (ED) is an extremely useful tool in understanding the basic properties of spin systems (especially in 2D), and it may be applied in the presence of large frustration. However, it is limited by the size of lattice that one may consider, typically up to a maximum of about 40 sites for spin-half systems, even using intensive computational methods. Because it may be applied only to smaller lattices, its accuracy is less than QMC.

Two other approximate methods that have proven very useful are cumulant series expansions and the CCM. Both methods provide results in the presence of frustration and for lattices of arbitrary spatial dimensionality. Although their accuracy tends to be slightly less than results of QMC, they may be applied to high orders of approximation via computational approaches. Indeed, computational methods are crucial to the application of all of these methods to high accuracy. Spin-wave theory is another method that also often provides excellent and complementary results. We have seen how a general consensus may be formed about the basic behaviour of a quantum spin model by applying of a broad range of approximate techniques, as for the Archimedean lattices. This is particularly important for those cases where the QMC or DMRG methods do less well, i.e., in the presence of frustration and for higher spatial dimensions, respectively.

Finally, we see also that the basic building blocks of unit cell, Bravais lattice, and bonds/interactions in the Hamiltonian placed on the lattice gives us a broad canvas to work with. For example, we may form models that interpolate between different lattices (and even different spatial dimensions) by varying the strengths of various bonds that have been carefully placed with respect to the underlying lattice. Hence, the number of possible such quantum spin systems is enormous.

Furthermore, the development in the number and complexity of these theoretical models is often driven by the magnetic materials studied in experiment.

Many interesting quantum phenomena that have no classical counterpart also may occur; ranging from spin plateau, valence-bond crystals, and spin liquids. We believe that the field of quantum spin systems will continue to provide many interesting challenges and new surprises well into the future.

References

1. Diep, T.H.: Frustrated Spin Systems. World Scientific, Singapore (2004)
2. Schollwöck, U., Richter, J., Farnell, D.J.J., Bishop, R.F. (eds.): Quantum Magnetism. Lecture Notes in Physics, vol. 645. Springer, Berlin (2004)
3. Mattis, D.C.: The Theory of Magnetism I, Springer, Berlin (1981, 1988).
4. Caspers, W.J.: Spin Systems. World Scientific, Singapore (1989)
5. Läuchli, A.M.: Quantum Magnetism and Strongly Correlated Electrons in Low Dimensions. Swiss Federal Institute of Technology, Zurich (2002)
6. Majumdar, C.K., Ghosh, D.K.: J. Math. Phys. **10**, 1388 (1969)
7. Majumdar, C.K., Ghosh, D.K.: J. Math. Phys. **10**, 1399 (1969)
8. Tonegawa, T., Harada, I.: J. Phys. Soc. Jpn. **56**, 2153 (1987)
9. Nomura, K., Okamoto, K.: Phys. Lett. **169A**, 433 (1992)
10. Nomura, K., Okamoto, K.: J. Phys. Soc. Jpn **62**, 1123 (1993)
11. Nomura, K., Okamoto, K.: J. Phys. A. Math. Gen. **27**, 5773 (1994)
12. Chitra, R., Pati, S., Krishnamurthy, H.R., Sen, D., Ramasesha, S.: Phys. Rev. B **52**, 6581 (1995)
13. White, S.R., Affleck, I.: Phys. Rev. B **54**, 9862 (1996)
14. Mikeska, H.-J., Kolezhuk, A.K.: One-dimensional magnetism. In: Schollwöck, U., Richter, J., Farnell, D.J.J., Bishop, R.F. (eds.) Quantum Magnetism. Lecture Notes in Physics, vol. 645, pp. 1–83. Springer, Berlin (2004)
15. Aligia, A.A., Batista, C.D., Eßler, F.H.L.: Phys. Rev. B **62**, 3259 (2000)
16. Zinke, R., Drechsler, S.-L., Richter, J.: Phys. Rev. B **79**, 094425 (2009)
17. Farnell, D.J.J., Parkinson, J.B.: J. Phys. Condens. Matter **6**, 5521 (1994)
18. Bursill, R., Gehring, G.A., Farnell, D.J.J., Parkinson, J.B., Xiang, T., Zeng, C.: J. Phys. Condens. Matter **7**, 8605 (1995)
19. Farnell, D.J.J., Richter, J., Zinke, R., Bishop, R.F.: J. Stat. Phys. **135**, 175 (2009)
20. White, S.R., Huse, D.A.: Phys. Rev. B **48**, 3844 (1993)
21. The exact diagonalisations *SpinPack* code of J. Schulenburg is available under GPL licence at: http://www.ovgu.de/jschulen/spin/
22. Farnell, D.J.J., Gernoth, K.A., Bishop, R.F.: J. Stat. Phys. **108**, 401 (2002)
23. Haldane, F.D.M.: Phys. Lett. **93A**, 464–468 (1983)
24. Haldane, F.D.M.: Phys. Rev. Lett. **50**, 1153–1156 (1983)
25. Nomura, K.: Phys. Rev. B **40**, 9142 (1989)
26. Takhtajan, L.A.: Phys. Lett. **87 A**, 479–482 (1982)
27. Babujian, H.M.: Phys. Lett. **90 A**, 479–482 (1982)
28. Lai, C.K.: J. Math. Phys. **15**, 1675–1676 (1974)
29. Sutherland, B.: Phys. Rev. B **12**, 3795–3805 (1975)
30. Parkinson, J.B.: J. Phys. C Solid State Phys. **20**, L1029–L1032 (1987)
31. Parkinson, J.B.: J. Phys. C Solid State Phys. **21**, 3793–3806 (1988)
32. Barber, M.N., Bachelor, M.T.: Phys. Rev. B **40**, 4621–4626 (1989)
33. Klümper, A.: Europhys. Lett. **9**, 815–820 (1989)
34. Klümper, A.: J. Phys. A Math. Gen. **23**, 809–823 (1990)

35. Affleck, I., Kennedy, T., Lieb, E.H., Tasaki, H.: Phys. Rev. Lett. **59**, 799–802 (1987)
36. Affleck, I., Kennedy, T., Lieb, E.H., Tasaki, H.: Commun. Math. Phys. **115**, 477–528 (1988)
37. Chubukov, A.V.: J. Phys.: Condens. Matter **2**, 1593–1608 (1990)
38. Chubukov, A.V.: Phys. Rev. B **43**, 3337–3344 (1991)
39. Fáth, G., Sólyom, J.: Phys. Rev. B **44**, 11836–11844 (1991)
40. Fáth, G., Sólyom, J.: Phys. Rev. B **47**, 872–881 (1994)
41. Xiang, T., Gehring, G.: Phys. Rev. B **48**, 303–310 (1993)
42. Manousakis, E.: Rev. Mod. Phys. **63**, 1 (1991)
43. Sandvik, A.W.: Phys. Rev. B **56**, 11678 (1997)
44. Hamer, C.J., Weihong, Z., Arndt, P.: Phys. Rev. B **46**, 6276 (1992)
45. Weihong, Z., Hamer, C.J.: Phys. Rev. B **43**, 8321 (1991)
46. Zeng, W., Oitmaa, J., Hamer, C.J.: Phys. Rev. B **43**, 8321 (1991)
47. Reger, J.D., Riera, J.A., Young, A.P.: J. Phys. Condens. Matter **1**, 1855 (1989)
48. Weihong, Z., Oitmaa, J., Hamer, C.J.: Phys. Rev. B **44**, 11869 (1991)
49. Oitmaa, J., Hamer, C.J., Weihong, Z.: Phys. Rev. B **45**, 9834 (1992)
50. Farnell, D.J.J., Bishop, R.F.: Int. J. Mod. Phys. B **22**, 3369 (2008)
51. Troyer, M., Kontani, H., Ueda, K.: Phys. Rev. Lett. **76**, 3822 (1996)
52. Ueda, K., Kontani, H., Sigrist, M., Lee, P.A.: Phys. Rev. Lett. **76**, 1932 (1996)
53. Weihong, Z., Gelfand, M.P., Singh, R.R.P., Oitmaa, J., Hamer, C.J.: Phys. Rev. B 55, 11377 (1997)
54. Farnell, D.J.J., Schulenburg, J., Richter, J., Gernoth, K.A.: Phys. Rev. B **72**, 172408 (2005)
55. Albrecht, M., Mila, F.: Europhys. Lett. **34**, 145 (1996).
56. Weihong, Z., Hamer, C.J., Oitmaa, J.: Phys. Rev. B **60**, 6608 (1999).
57. Miyake, S.J.: J. Phys. Soc. Jpn. **61**, 983 (1992)
58. Singh, R.P., Huse, D.A.: Phys. Rev. Lett. **68**, 1766 (1992)
59. Schmallfuß, D., Tomczak, P., Schulenburg, J., Richter, J.: Phys. Rev. B **63**, 22405 (2002)
60. Harris, A.B., Kallin, C., Berlinsky, A.J.: Phys. Rev. B **45**, 2899 (1993).
61. Farnell, D.J.J., Bishop, R.F., Gernoth, K.A.: Phys. Rev. B **63**, 220402(R) (2001)
62. Collins, M.F., Petrenko, O.A.: Can. J. Phys. **75**, 655 (1997)
63. Honecker, A.: J. Phys. Condens. Matter **11**, 4697 (1999)
64. Cabra, D.C., Grynberg, M.D., Honecker, A., Pujol, P.: In: Hernández, S., Clark, J.W. (eds.) Magnetization Plateaux in Quasi-One-Dimensional Strongly Correlated Electron Systems. Condens. Matter Theor., vol. 16, p. 17. Nova Science Publishers, New York, NY (2001); arXiv:cond-mat/0010376v1
65. Lhuillier, C., Misguich, G.: In: Berthier, C., Lévy, L.P., Martinez, G. (eds.) Frustrated Quantum Magnets. High Magnetic Fields. Lecture Notes in Physics, vol. 595, p. 161. Springer, Berlin (2002)
66. Honecker, A., Schulenburg, J., Richter, J.: J. Phys. Condens. Matter **16**, S749 (2004)
67. Nishimori, H., Miyashita, S.: J. Phys. Soc. Jpn. **55**, 4448 (1986)
68. Chubukov, A.V., Golosov, D.I.: J. Phys. Condens. Matter **3**, 69 (1991)
69. Alicea, J., Chubukov, A.V., Starykh, O.A.: Phys. Rev. Lett. 102, 137201 (2009)
70. Chubukov, A.V., Sachdev, S., Senthil, T.: J. Phys. Condens. Matter **6**, 8891 (1994)
71. Trumper, A.E., Capriotti, L., Sorella, S.: Phys. Rev. B **61**, 11529 (2000)
72. Schulenburg, J., Richter, J.: Phys. Rev. **65**, 054420 (2002)
73. Ono, T., Tanaka, H., Aruga Katori, H., Ishikawa, F., Mitamura, H., Goto, T.: Phys. Rev. B **67**, 104431 (2003)
74. Farnell, D.J.J., Richter, J., Zinke, R., J. Phys. Condens. Matter 21, 406002 (2009)
75. Fortune, N.A., Hannahs, S.T., Yoshida, Y., Sherline, T.E., Ono, T., Tanaka, H., Takano, Y.: arXiv:0812.2077v1
76. Bishop, R.F., Farnell, D.J.J., Parkinson, J.B.: Phys. Rev. B **58**, 6394 (1998)
77. Bishop, R.F., Li, P.H.Y., Darradi, R., Schulenburg, J., Richter, J.: Phys. Rev. B **78**, 054412 (2008)
78. Bishop, R.F., Li, P.H.Y., Darradi, R., Richter, J.: J. Phys. Condens. Matter **20**, 255251 (2008)
79. Darradi, R., Richter, J., Schulenburg, J.: J. Phys. Conf. Ser. **145**, 012049 (2009)

80. Darradi, R., Derzhko, O., Zinke, R., Schulenburg, J., Krüger, S.E., Richter, J.: Phys. Rev. B **78**, 214415 (2008)
81. Oitmaa, J., Weihong, Z.: Phys. Rev. B **54**, 3022 (1996)
82. Singh, R.R.P., Weihong, Z., Hamer, C.J., Oitmaa, J.: Phys. Rev. B **60**, 7278 (1999)
83. Kotov, V.N., Oitmaa, J., Sushkov, O., Weihong, Z.: Philos. Mag. B **80**, 1483 (2000)
84. Sirker, J., Weihong, Z., Sushkov, O.P., Oitmaa, J.: Phys. Rev. B **73**, 184420 (2006)
85. Pardini, T., Singh, R.R.P.: Phys. Rev. B **79**, 094413 (2009)
86. Dagotto, E., Moreo, A.: Phys. Rev. Lett. **63**, 2148 (1989)
87. Dagotto, E., Moreo, A.: Phys. Rev. B **39**, 4744(R) (1989)
88. Schulz, H.J., Ziman, T.A.L.: Europhys. Lett. **18**, 355 (1992)
89. Schulz, H.J., Ziman, T.A.L., Poilblanc, D.: J. Phys. I **6**, 675 (1996)
90. Richter, J., Schulenburg, J.: arXiv:0909.3723v1
91. Isaev, L., Ortiz, G., Dukelsky, J.: Phys. Rev. B **79**, 024409 (2009)
92. Chandra, P., Doucot, B.: Phys. Rev. B **38**, 9335 (1988)
93. Capriotti, L., Becca, F., Parola, A., Sorella, S.: Phys. Rev. Lett. **87**, 097201 (2001)
94. Shastry, B.S., Sutherland, B.: Physica B **108**, 1069 (1981)
95. Weihong, Z., Oitmaa, J., Hamer, C.J.: Phys. Rev. B **65**, 014408 (2002)
96. Läuchli, A., Wessel, S., Sigrist, M.: Phys. Rev. B **66**, 014401 (2002)
97. Müller-Hartmann, E., Singh, R.R.P., Knetter, C., Uhrig, G.S.: Phys. Rev. Lett. **84**, 1808 (2000)
98. Koga, A., Kawakami, N.: Phys. Rev. Lett. **84**, 4461 (2000)
99. Al Hajii, M., Guihery, N., Malrieu, J.-P., Bouquillon, B.: Eur. Phys. J. B **41**, 11 (2004)
100. Chung, C.H., Marston, J.B., Sachdev, S.: Phys. Rev. B **64**, 134407 (2001)
101. Miyahara, S., Ueda, K.: J. Phys.: Condens. Matter **15**, R327 (2003)
102. Darradi, R., Richter, J., Farnell, D.J.J.: Phys. Rev. B. **72**, 104425 (2005)
103. Farnell, D.J.J., Richter, J., Zinke, R., Bishop, R.F.: J. Stat. Phys. **135**, 175 (2009)

Index

Parkinson, J.B., Farnell, D.J.J.: *Index*. Lect. Notes Phys. **816**, 153–154 (2010)
DOI 10.1007/978-3-642-11914-9